Principles of
Applied Geophysics

Principles of
Applied Geophysics

D. S. Parasnis

Professor of Applied Geophysics
University of Luleå, Luleå, Sweden
Fellow of the Royal Swedish Academy
of Engineering Sciences

THIRD EDITION

LONDON   NEW YORK

CHAPMAN AND HALL

First published 1962
by Methuen & Co. Ltd.
Reprinted once
Reprint 1971 published by
Chapman and Hall Ltd.,
11 New Fetter Lane, London EC4P 4EE
Second edition 1972
Reprinted once
Third edition 1979
Reprinted 1982
Published in the U.S.A. by Chapman and Hall,
in association with Methuen, Inc.,
733 Third Avenue, New York NY 10017

© *1962, 1972, 1979 D. S. Parasnis*

Printed in Great Britain by
Richard Clay (*The Chaucer Press*) *Ltd., Bungay, Suffolk*

ISBN 0 412 15140 5 (cased edition)
ISBN 0 412 15810 8 (Science Paperback edition)

Preface to the third edition

The welcome accorded to the first two editions of this book has been most encouraging. The object of the third edition continues to be to give a brief but fairly comprehensive survey of the methods of applied geophysics including some of the modern interpretation techniques. The general approach and plan of the previous editions are preserved, but in bringing the book up to date some changes have been made to which I would like to draw the reader's special attention.

SI units are strictly adhered to except in six illustrative figures reproduced from older literature and left intact to save some extensive redraughting. Following the recommendation of the International Union of Geodesy and Geophysics, the magnetic field measured in geophysical work is labelled here as flux density (tesla). Consequently, the symbols H, Z and T commonly used in geomagnetic work should stand for flux density. In the Maxwellian theory of electromagnetism the symbol H stands, by convention, for a magnetizing force $(A\ m^{-1})$ and a discerning reader will at once sense a source of confusion.

This source of confusion is avoided in the present edition by employing the symbols B_h, B_z and B_t instead of H, Z and T. The latter set is employed for the corresponding *magnetizing forces* of the earth's field. I hope this notation will gain general acceptance because it so easily dispenses with an ambiguity that otherwise tends to lead to unnecessary confusion of units and dimensions in geomagnetism.

The reader will also notice that the entire magnetic theory is logically developed here without recourse to or, in fact, without any reference at all to the concept of a magnetic pole. This is highly desirable for a consistent adoption of SI.

In spite of its shortness the present edition, like the two

previous ones, contains a number of topics not to be found even in much bulkier current texts in English devoted to the subject as a whole. Among these can be mentioned a generalized treatment of the magnetic anomalies of thick and thin sheets, a discussion of Green's theorem in potential theory, the kernel function theory of resistivity sounding, the powerful methods of Ghosh, Inman and Johansen in VES interpretation, Lee's method for dipping discontinuities, Dar Zarrouk curves, a discussion of near and far regions of the electromagnetic field, design principles of electro-magnetic sensors, radioactive density determinations, statistical considerations in optimum line spacing and a number of others.

The induced polarization method is now devoted a chapter of its own while the material in the separate chapter on airborne methods in the previous editions is now incorporated under the respective methods. The detailed derivation of some of the mathematical formulae is given in the appendices, or in special sections, in order that the main text should flow as unhindered as possible. Numbers in square brackets in the text denote the literature cited in the list of references at the end of the book. The examples in Fig. 44 and Table 8 are due to Hans-Kurt Johansen of the Department of Applied Geophysics, University of Luleå.

Finally, I should like to thank Irene Lundmark, Department of Applied Geophysics, University of Luleå, for her very efficient help in the preparation of the typescript of this edition.

Luleå, Sweden   D. S. Parasnis

Contents

x / *Contents*

1 Introduction

Geophysics is the application of the principles of physics to the study of the earth. Strictly speaking the subject includes meteorology, atmospheric electricity, or ionosphere physics; but in this monograph the word geophysics will be used in the more restricted sense, namely the physics of the body of the earth. The aim of pure geophysics is to deduce the physical properties of the earth and its internal constitution from the physical phenomena associated with it, for instance the geomagnetic field, the heat flow, the propagation of seismic waves, the force of gravity, etc. On the other hand, the object of applied geophysics, with which this monograph is concerned, is to investigate specific, relatively small-scale and shallow features which are presumed to exist within the earth's crust. Among such features may be mentioned synclines and anticlines, geological faults, salt domes, undulations of the crystalline bedrock under a cover of moraine, ore bodies, clay deposits and so on. It is now common knowledge that the investigation of such features very often has a bearing on practical problems of oil prospecting, the location of water-bearing strata, mineral exploration, highway construction and civil engineering. Often, the application of physics, in combination with geological information, is the only satisfactory way towards a solution of these problems.

The geophysical methods used in investigating the shallow features of the earth's crust vary in accordance with the physical properties of the rocks – the last word is used in the widest sense – of which these features are composed, but broadly speaking they fall into four classes. On the one hand are the *static methods* in which the distortions of a static physical field are detected and measured accurately in order to delineate the features producing them. The static field may be a natural field

1

like the geomagnetic, the gravitational or the thermal gradient field, or it may be an artificially applied field like an electric potential gradient. On the other hand, we have the *dynamic methods* in which signals are sent into the ground, the returning signals are detected, and their strengths and times of arrival are measured at suitable points. In the dynamic methods the dimension of time always appears, in the appropriate field equations, either directly as the time of wave arrival, as in the seismic method, or indirectly as the frequency or phase difference, as in the electromagnetic method. There is a further, now considerably important, class of methods which lie in between the two just mentioned. These will be called *relaxation methods*. Their feature is that the dimension of time appears in them as the time needed for a disturbed medium to return to its normal state. This class includes the overvoltage or induced polarization methods. Finally there are what we may call *integrated effect methods*, in which the detected signals are statistical averages over a given area or within a given volume. The methods using radioactivity fall in this class.

The classification of geophysical methods into ground, airborne or borehole methods refers only to the operational procedure. It has no physical significance. Many ground methods can be used in the air, under water or in boreholes as well.

The magnetic, electromagnetic and radioactive methods have been adapted to geophysical measurements from the air. Airborne work has certain advantages. Firstly, on account of the high speed of operations an aerial survey is many times cheaper than an equivalent ground survey, provided the area surveyed is sufficiently large, and secondly, measurements can be made over mountains, jungles, swamps, lakes, glaciers and other terrains which may be inaccessible or difficult for ground surveying parties.

Compared with ground work, airborne measurements imply a decrease in resolution which means that adjacent geophysical indications tend to merge into one another giving the impression of only one indication. Besides, there is often considerable uncertainty about the position of airborne indications so that they must be confirmed on the ground before undertaking further work such as drilling.

In a sense, applied geophysics, excepting the seismic methods, is predominantly a science suited to flat or gently undulating terrain where the overburden is relatively thin. The reason is that whenever the relief is violent, the data of geophysical methods need corrections which are frequently such as to render their interpretation uncertain. On the other hand, when the overburden is too thick the effects produced by the features concealed under it generally lie within the errors of measurement and are difficult to ascertain. There is, however, no general rule as to the suitability of any terrain to geophysical methods and every case must be considered carefully on its own merits.

The various methods of applied geophysics will be dealt with in turn in the following chapters.

2 Magnetic methods

2.1 **Short history**

It was in the year 1600 when William Gilbert, Physician to Queen Elizabeth I, published his book *De Magnete* (abbreviated title), that the concept arose of a general geomagnetic field with a definite orientation at each point on the surface of the earth. In its wake, observations of the local anomalies in the orientation of the geomagnetic field were used in Sweden for iron-ore prospecting, for the first time probably as early as 1640 and regularly by the end of that century. They constitute the first systematic utilization of a physical property for locating specific, small-scale features within the earth's crust. Two centuries later, in 1870, Thalén constructed his magnetometer for comparatively rapid and accurate determinations of the horizontal force, the vertical force and the declination, by the familiar sine and tangent methods used in elementary physics courses. This and its somewhat simplified modification due to Tiberg were in widespread use, especially in Sweden, as a tool for prospecting surveys for more than the following half a century.

The large-scale use of magnetic measurements for investigations of geological structures, other than those associated with magnetic ore, did not however begin seriously until 1915, when Adolf Schmidt constructed his precision vertical field balance using a magnetic needle swinging on an agate knife edge. Since then magnetic observations have been successfully employed, not only in the search for magnetite ore, but also in locating buried hills, geological faults, intrusions of igneous rocks, salt domes associated with oil fields, concealed meteorites and buried magnetic objects such as pipe-lines.

2.2 The static magnetic field

2.2.1 Fundamental definitions

A magnetizing force H (A/m) at a point in any medium gives rise to a flux density $B = \mu H$ (Vs/m^2, tesla) at that point. μ is known as the absolute permeability and evidently has the dimensions ohmsecond per metre. In vacuum the same H will create a flux density $\mu_0 H$ where μ_0 is the permeability of vacuum. In the International System of Units (SI) the unit of current is (in effect) defined so that the value of μ_0 turns out to be $4\pi \times 10^{-7}$ Ωs/m.

Writing $\mu = \mu_r \mu_0$ we see that

$$B = \mu_r \mu_0 H$$
$$= \mu_0 H + \mu_0 (\mu_r - 1)H$$
$$= \mu_0 H + \mu_0 \kappa H \quad (\kappa = \mu_r - 1) \tag{2.1}$$

From the last line of (2.1) it is obvious that to obtain in vacuum a flux density equal to the density in the medium under consideration, we would need an additional magnetizing force κH. This additional magnetizing force that may be said to present at points of space occupied by a medium is called the *intensity of magnetization M* (A/m), while μ_r and κ are known respectively as the relative permeability and the susceptibility.

In this book we shall adopt the rationalized system of units in which the measures of M and H are equal. In the so-called unrationalized system these measures differ (through definition) by a factor of 4π and the value of κ in it is smaller by this factor than the value in the rationalized system. In either system, however, κ is a pure number, notwithstanding the physically superfluous, and hence misleading, practice in much current geophysical literature of implying a 'dimension' for κ (e.g. e.m.u./oersted, e.m.u./cm^3 etc.). Evidently μ_r is also a pure number.

Since both B and H are vectors we have, in general,

$$\mathbf{B} = \mu_0 (\mathbf{H} + \mathbf{M})$$

and for the x, y, z components in an orthogonal coordinate system

$$B_x = \mu_0 (H_x + M_x), \text{ etc.} \tag{2.2}$$

Like the mechanical force (joule per metre = newton) or the electrical force (volt per metre), a static magnetizing force (ampere per metre) is also postulated to be derivable from a corresponding (scalar) potential ϕ (ampere) so that

$$H_x = -\frac{\partial\phi}{\partial x}, \quad H_y = -\frac{\partial\phi}{\partial y}, \quad H_z = -\frac{\partial\phi}{\partial z} \tag{2.3}$$

No sources or sinks of magnetic flux have ever been found to exist. The 'pole' of a magnet is not a source or a sink of flux. If it were, we would be able to isolate it as such by cutting the magnet, but no subdivision of a magnet, however fine, succeeds in this respect.

It is easy to show that in a field of flux that contains no sources or sinks of flux, the divergence of the flux density vector is zero. For the magnetic flux we have thus

$$\mathrm{div}\,\mathbf{B} = \frac{\partial B_x}{\partial x} + \frac{\partial B_y}{\partial y} + \frac{\partial B_z}{\partial z} = 0 \tag{2.4}$$

From the sets of Equations (2.2) and (2.3) in combination with (2.4) it follows immediately that

$$\nabla^2 \phi = \mathrm{div}\,\mathbf{M} \tag{2.5}$$

It is shown in Appendix 1 that if M_x, M_y, M_z are the components of $\mathbf{M}$ at any point $P(x, y, z)$ the potential ϕ at a point $Q(\xi, \eta, \zeta)$ can be calculated from

$$\phi = \frac{1}{4\pi} \iiint\limits_{R} \left(M_x \frac{\partial}{\partial x}\left(\frac{1}{r}\right) + M_y \frac{\partial}{\partial y}\left(\frac{1}{r}\right) + M_z \frac{\partial}{\partial z}\left(\frac{1}{r}\right) \right) dv \tag{2.6}$$

where dv is a volume element at P and

$$r^2 = (x - \xi)^2 + (y - \eta)^2 + (z - \zeta)^2$$

and the integrations are to be extended over all the regions R within which $\mathbf{M} \neq 0$. Strictly speaking the integration is throughout all space but the integrals vanish in regions where $\mathbf{M} = 0$ (e.g. vacuum or, for all practical purposes in geophysics, air). H_x, H_y, H_z at Q can be found by differentiating ϕ with respect to ξ, η, ζ and B_x, B_y, B_z are then given simply by $\mathbf{B} = \mu_r\mu_0\mathbf{H}$. This is the basis, in principle, of calculating the magnetic effect of any body.

2.2.2 Dipole potential

If $\mathbf{M} \neq 0$ within an infinitesimal volume Δv and 0 elsewhere, the derivatives in (2.6) may be considered to be constant throughout

Δv and

$$\phi = \frac{1}{4\pi} \left[\frac{\partial}{\partial x} \left(\frac{1}{r} \right) \int\limits_{\Delta v} M_x \, dv + \frac{\partial}{\partial y} \left(\frac{1}{r} \right) \int\limits_{\Delta v} M_y \, dv \right.$$

$$\left. + \frac{\partial}{\partial z} \left(\frac{1}{r} \right) \int\limits_{\Delta v} M_z \, dv \right]$$

$$= \frac{1}{4\pi} \left[m_x \frac{\partial}{\partial x} \left(\frac{1}{r} \right) + m_y \frac{\partial}{\partial y} \left(\frac{1}{r} \right) + m_z \frac{\partial}{\partial z} \left(\frac{1}{r} \right) \right] \qquad (2.7)$$

The integrals of M_x, M_y, M_z over the small volume Δv are the magnetic moments (Am^2) in the x, y, z directions. In the limit $\Delta v \rightarrow 0$ Equation (2.7) gives the potential of a point dipole whose moment **m** has the components m_x, m_y, m_z in the x, y, z directions.

In the case of a point dipole there is no loss of generality in rotating the coordinate axes so that the components of **m** along the new y and z directions vanish, in which case

$$\phi_{\text{dipole}} = \frac{m}{4\pi} \frac{\partial}{\partial x} \left(\frac{1}{r} \right)$$

$$= - \frac{m}{4\pi} \frac{\partial}{\partial \xi} \left(\frac{1}{r} \right) = \frac{m}{4\pi} \frac{1}{r^2} \frac{\xi - x}{r}$$

$$= \frac{m}{4\pi} \frac{1}{r^2} \cos \theta \qquad (2.8)$$

where θ is the angle between the dipole axis and the line joining the dipole at P to the point Q.

2.2.3 Magnetic moment

For a single body of volume V where the surrounding medium is vacuum, Equation (2.6) may be written using the mean value theorem of integral calculus as:

$$\phi = \frac{1}{4\pi} \left[\left\{ \frac{\partial}{\partial x} \left(\frac{1}{r} \right) \right\}_1 m_x + \left\{ \frac{\partial}{\partial y} \left(\frac{1}{r} \right) \right\}_2 m_y + \left\{ \frac{\partial}{\partial z} \left(\frac{1}{r} \right) \right\}_3 m_z \right]$$

where $m_x = \iiint_V M_x \, dv$ etc., and 1, 2, 3 denote three definite (but undetermined) points in the interior of the body. *Provided the magnetization is uniform* we have $m_x = M_x V$, $m_y = M_y V$, $m_z = M_z V$ so that in this case the intensity of magnetization is seen to be the magnetic moment per unit volume.

For non-uniformly magnetized bodies m_x/V etc. is the average intensity of magnetization.

2.3 Magnetic properties of rocks

The magnetic method of applied geophysics depends upon measuring accurately the anomalies of the local geomagnetic field produced by the variations in the intensity of magnetization in rock formations. The magnetization of rocks is partly due to induction by the magnetizing force associated with the earth's field and partly to their permanent (remanent) magnetization. The induced intensity depends primarily upon the magnetic suscepti-bility as well as the magnetizing force, and the permanent intensity upon the geological history of the rock. Research in the permanent intensity of rocks, especially since 1950, has given rise to the subject of palaeomagnetism.

2.3.1 Susceptibility of rocks

In accordance with the general classification used in modern physics, rocks (like all substances) fall into three categories, namely diamagnetic, paramagnetic and ferromagnetic. The last named category is further subdivided into the truly ferromagnetic, the antiferromagnetic and the ferrimagnetic substances. These terms are briefly explained below.

The intensity of magnetization, M_i, induced in isotropic substances due to a magnetizing force H (A/m) can be written as:

$$M_i = \kappa H \tag{2.9}$$

To be more general, both M_i and H are vectors and κ a tensor so that a force in say direction 1 produces an induced intensity in this direction as well as in the two mutually orthogonal directions and so on for the other directions. The susceptibility tensor has six independent components which can be experimentally determined for a given rock sample.

In a diamagnetic substance κ is negative so that the induced intensity is in a direction opposite to the magnetizing force. The origin of diamagnetism lies in the motion of an electron round a nucleus. This motion constitutes a miniature plane current circuit and is characterized by a magnetic moment vector as well as an angular momentum vector, both at right angles to the plane of the electron's motion.

An impressed magnetizing force will tend to turn the magnetic moment in the direction of the force and a mechanical torque will act on the orbital plane. The orbit reacts in a way known, sufficiently well for our purpose, from the behaviour of a top subject to a torque tending to turn its angular momentum. The result, as in the case of the top, is that the angular momentum vector and hence also the magnetic moment vector begin to precess round the magnetizing force. This is known as Larmor precession. The additional periodic motion of the electron due to Larmor precession is such as to produce a magnetic moment opposite in direction to the applied field.

It will be realized that there is a diamagnetic effect in all substances including the 'typical' ferromagnetics like iron, cobalt and nickel. But net diamagnetism only appears if the magnetic moments of atoms are zero in the absence of an external magnetizing force, as is the case for atoms or ions having closed electronic shells. There are many rocks and minerals which show net diamagnetism. Chief among them are quartz, marble, graphite, rock salt, and anhydrite (gypsum).

The susceptibility of paramagnetic substances is positive and decreases inversely as the absolute temperature (Curie–Weiss law). Paramagnetism makes its appearance when the atoms or molecules of a substance have a magnetic moment in the absence of a field and the magnetic interaction between the atoms is weak. Normally the moments are distributed randomly, but on the application of the field they tend to align themselves in the direction of the field, the tendency being resisted by thermal agitation. The paramagnetism of elements is mainly due to the unbalanced spin magnetic moments of the electrons in unfilled shells, like the 3d-shells of the elements from Sc to Mn. Many rocks are reported to be paramagnetic, for instance gneisses, pegmatites, dolomites, syenites, etc. However, it seems certain that their paramagnetism is not intrinsic but is a manifestation of a weak ferrimagnetism due to varying amounts of magnetite or ilmenite, or an antiferromagnetism due to minerals like haematite, manganese dioxide, etc.

In ferromagnetic materials the atoms have a magnetic moment and the interaction between neighbouring atoms is so strong that the moments of all atoms within a region, called a domain, align themselves in the same direction even in the absence of an external field. In Fe, Co and Ni this interaction takes place between the uncompensated spins in the unfilled 3d-shells of the atoms. A state of spontaneous magnetization can therefore exist consisting of an orderly arrangement of the magnetic moments of all atoms. Typical of the ferromagnetics are their hysteresis loops and their large susceptibilities which depend upon the magnetizing force. Ferromagnetism disappears above a temperature known as the Curie temperature. There are no truly ferromagnetic rocks or rock materials.

There exist substances in which the susceptibility has an order of magnitude characteristic of a paramagnetic (10^{-5}) but is not inversely proportional to the temperature. Instead it first increases with temperature, reaches a maximum at a certain temperature, also called the Curie point or the λ-point, and decreases thereafter according to the Curie–Weiss law. In these substances the low magnetic susceptibility below the λ-point can be explained by assuming an *ordered* state of atoms such that the magnetic moments of neighbouring atoms are equal but directed antiparallel to each other. Thus the two ordered sub-lattices, each reminiscent of the state in a ferromagnetic, cancel each other and their net magnetic moment is zero. This state is called antiferromagnetism and can be confirmed by neutron diffraction studies. Of the rock-forming minerals, haematite (Fe_2O_3) is the most important antiferromagnetic (λ-point 675° C).

Among the antiferromagnetic substances there is a class in which, to put it simply, two sub-lattices with metallic ions having magnetic moments are ordered antiparallel as above, but in which the moments of the lattices are unequal, giving rise to a net magnetic moment in the absence of a field. Such substances are called ferrimagnetic. Practically all the constituents giving a high magnetization to rocks are ferrimagnetic, chief among them being magnetite (Fe_3O_4), titanomagnetite ($FeO(Fe, Ti)_2O_3$) and ilmenite ($FeTiO_3$). Spontaneous magnetization and a relatively high susceptibility can also exist in an antiferromagnetic if statistically systematic defects are present, as is believed to be the case for pyrrhotite (FeS). The temperature dependence of ferrimagnetics is complex, there being theoretically several possibilities.

The susceptibility of rocks is almost entirely controlled by the amount of ferrimagnetic minerals in them, their grain size, mode of distribution, etc. and is extremely variable. The values listed in

Table 1   Susceptibilities $\times 10^6$ (rationalized system)

Graphite	−100	Gabbro	3800−90 000
Quartz	−15.1	Dolomite	
Anhydrite	−14.1	(impure)	20 000
Rock salt	−10.3	Pyrite	
Marble	−9.4	(pure)	35−60
Dolomite		Pyrite	
(pure)	−12.5 − +44	(ore)	100−5000
Granite		Pyrrhotite	10^3−10^5
(without magnetite)	10−65	Haematite (ore)	420−10 000
Granite		Ilmenite (ore)	3×10^5−4×10^6
(with magnetite)	25−50 000	Magnetite (ore)	7×10^4−14×10^6
Basalt	1500−25 000	Magnetite (pure)	1.5×10^7
Pegmatite	3000−75 000		

To convert the above values to the unrationalized system divide by 4π.

Table 1 should nevertheless serve to give a rough idea. Various attempts have been made to represent the dependence of susceptibility on the content of ferrimagnetics, but no simple universally valid relation exists. For particular groups of rocks or for particular ranges of susceptibility a statistically significant correlation can generally be found between the amount of $Fe_3 O_4$, and the susceptibility (Fig. 1). However, the scatter is usually such that a prediction based on such correlations must be used with caution.

It will be seen that generally speaking only small susceptibility differences ($\Delta\kappa$) will be encountered between rock formations. The maximum $\Delta\kappa$ (when a deposit of high grade magnetite ore is present) is of the order of ten. As we shall see later, if two very extensive and thick homogeneous formations are separated along a plane vertical contact the peak-to-peak change in the vertical magnetic field in traversing the contact is given by

$$\Delta B_z = \tfrac{1}{2}\Delta\kappa B_z \tag{2.10}$$

if B_z tesla (T) is the vertical magnetic field (flux density) of the earth. The practical limit of determining ΔB_z in most field surveys is about ·1 nanotesla (nT) and if B_z = 50 000 nT, substitution in the above equation shows that the practical limit of detecting a susceptibility difference between rock formations will be $\Delta\kappa \approx 4 \times 10^{-5}$. However, the lack of homogeneity in rocks and their impregnation by ferrimagnetic minerals produce random magnetic anomalies, the 'geologic noise', owing to which the limit

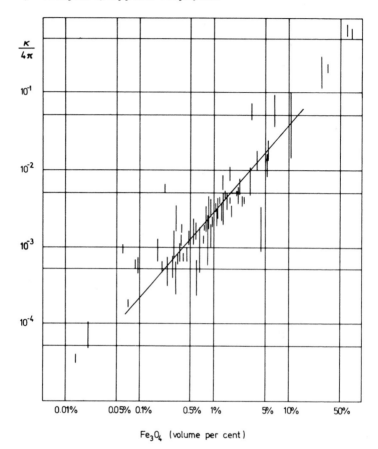

Fig. 1. Variation of susceptibility κ (rationalized system) with magnetite content. After Balsley and Buddington (*Economic Geology*, 1958).

in reality is some $10-100$ times larger. The maximum peak-to-peak change (over a magnetic deposit) will be found from (2.2) by putting $\Delta\kappa = 12$ to be about 300 000 nT. In these calculations, which are intended to illustrate the orders of magnitude involved, no account has been taken of the permanent magnetization of rocks.

2.3.2 Permanent magnetization of rocks

Researches in palaeomagnetism all over the world have confirmed that both igneous and sedimentary rocks possess permanent magnetization in varying degrees and that the phenomenon is a widespread one. Well-documented examples of rocks, igneous as well as sedimentary, occur in all parts of the world and of all geological ages, in which the permanent intensity is not only strong but has a direction completely different from, at times opposite to, the present direction of the geomagnetic field.

Various types of permanent magnetization of rocks are now recognized. One principal type, especially for igneous rocks, is thermoremanent magnetization (TRM) acquired in cooling from high temperatures. Its orientation reflects the orientation of the geomagnetic field prevalent at the time and place of formation. The predominant mechanism in the acquisition of TRM is the alignment of the domains in the ferrimagnetic constituents of the rocks. It is in this respect significant that the TRM of rocks disappears when they are heated above 600° C, which is approximately the Curie point of magnetite [1]. Other principal types of natural remanent magnetization (NRM) of rocks are: isothermal remanent magnetization (IRM) acquired at constant temperature on exposure to a magnetizing force for a short time; viscous (VRM) acquired as a cumulative effect after a long exposure in an ambient field, not necessarily at one and the same temperature; depositional or detrital (DRM) acquired by sediments as the constituent magnetic grains settle in water under the influence of the earth's field; chemical (CRM) acquired during growth or recrystallization of magnetite grains at temperatures far below Curie temperatures.

Some examples of the ratio $Q_n = M_n/(\kappa T)$, the Koenigsberger ratio, where M_n is the permanent intensity and T the total magnetizing force of the earth's field, are given in Table 2. In most igneous rocks like gabbro and basalts, the permanent intensity completely dominates the intensity induced by the earth's field. Hence magnetic interpretation, especially in areas where igneous rocks occur, must take into account the permanent intensity if a satisfactory geological picture is to be obtained. The same applies when the object of investigation is magnetite ore.

The resultant magnetization M of a rock can be described by the vector equation

$$\mathbf{M} = \mathbf{M_n} + \kappa \mathbf{T} \tag{2.11}$$

In the simplest cases when M and F are parallel and antiparallel and the rock is isotropic, an apparent susceptibility

$$\kappa' = \kappa \pm M_n/T \tag{2.12}$$

Table 2 Q_n for some rock specimens

Specimen	Locality	Q_n
Basalt	Mihare volcano, Japan	99–118
Gabbro	Cuillin Hills, Scotland	29
Gabbro	Småland, Sweden	9.5
Andesite	Taga, Japan	4.9
Granite	Madagascar	0.3–10
Quartz dolerite	Whin sill, England	2–2.9
Diabase	Astano Ticino, Switzerland	1.5
Tholeiite dikes	England	0.6–1.6
Dolerite	Sutherland, Scotland	0.48–0.51
Magnetite ore	Sweden	1–10
Manganese ore	India	1–5

Sediments generally low.

can be defined and used during interpretation in the equations specifying the anomaly of an anomalous feature. For the modification necessary in the general case reference may be made to a paper by Green [2]. Methods for determining the susceptibility and remanence of rocks are discussed in Section 2.13.

2.4 The geomagnetic field

In order to identify the anomalies in the earth's field it is clearly essential to know its undisturbed character. To a very close approximation the regular geomagnetic field can be represented formally as the field of a dipole situated at the centre of the earth with its magnetic moment pointing towards the earth's geographical south. Physically, the origin of the field seems to be a system of electric currents within the earth. The total geomagnetic flux density B_t (on the surface) has a magnitude of about 0.6×10^{-4} T. This flux density is the result of a magnetizing force T_0 existing at the point of observation of magnitude $0.6 \times 10^{-4}/(4\pi \times 10^{-7}) = 47.8$ A/m (cf. Section 2.2). At any point on the earth's surface the magnetic flux density vector (or the magnetic field, as we shall often call the flux density) is completely specified by its horizontal (B_h) and vertical (B_z) components and the declination (D), west or east of true north, of B_h. The direction of B_h is the local magnetic meridian. B_z is reckoned positive if it points downwards as in the northern hemisphere generally, and negative if it points upwards as in the southern hemisphere. The inclination (I) of B_h, which is of importance in

Table 3   Values of B_h, B_z, D (1975)

Place	Lat.	Long. East	B_h nT	B_z nT	D East
Chelyuskin	+77.7°	104.3°	3510	59130	16.9°
Sodankylä	+67.4	26.6	11950	50530	6.2
Hartland	+51.0	355.5	19210	43730	−8.5
San Miguel	+37.8	334.4	25540	38030	−13.3
Hyderabad	+17.4	78.6	39870	14920	−1.6
Tangerang	−6.2	106.6	37190	−23810	0.8
Apia	−13.8	188.2	34450	−20120	12.3
Mauritius*	−20.4	57.7	21720	−29030	−16.9
Gnangara	−31.8	116.0	23610	−53500	−3.2
Macquarie Is.	−54.5	159.0	12850	−63930	27.7
Scott Base	−77.8	166.8	10510	−67840	156.4

*Values of B_h, B_z, D for 1976.

the interpretation of magnetic anomalies, is given by $\tan^{-1}$ (B_z/B_h). The points on the earth at which $I = \pm 90°$ are called the magnetic north and south dip poles respectively. There may be any number of such points due to local disturbances, but apart from them there are two main north and south dip poles situated approximately at 72° N, 102° W and 68° S, 146° E. On account of the irregular part of the earth's field they do not correspond to the intersections of the axis of the imaginary dipole at the earth's centre with the surface which are at 79° N, 70° W and 79° S, 110° E. The latter are called the geomagnetic poles or axis poles. The imaginary line on the earth's surface passing through the points $I = 0$ is called the magnetic equator. North of it B_z is positive; south of it B_z is negative. The great circle on the earth's surface passing through the imaginary dipole and at right angles to the dipole axis is the geomagnetic equator.

The earth's field is not constant at any point on its surface but undergoes variations of different periods. From the standpoint of applied geophysics the most important are the diurnal variations and magnetic storms. Their disturbing effect must be suitably eliminated from magnetic survey observations. Values of B_h, B_z and D for the epoch 1975 at some selected places are given in Table 3.

2.5 Instruments of magnetic surveying

Magnetic measurements in applied geophysics are often carried out as relative determinations in which the values of one or more

Table 4 Variation of B_h, B_z, B_t at geomagnetic latitude of 50°

Variation	B_h	B_z	B_t
With height	−0.96 nT/100 m	−2.27 nT/100 m	−2.47 nT/100 m
With latitude towards N	−3.79 nT/km	6.35 nT/km	4.40 nT/km

elements of the magnetic field at any point are measured as differences from the values at a suitably chosen base point. Many modern instruments, however, read directly the absolute values of, for example, the total field or the vertical field. Usually the area of investigation is relatively small, say a few square kilometres, so that the normal geomagnetic field within it may be considered to be substantially constant and equal to that at the base point. If absolute measurements are made the base point field is usually subtracted from the observations and only the resulting anomalies are used for interpretation. In very large areas, say more than a hundered square kilometres, the variation of the normal field may be significant and should be corrected for. This may be advisable in a large-scale aerial survey. Table 4 shows the variations in the normal geomagnetic field with latitude and height above ground surface at the geomagnetic latitude of 50°.

A variety of instruments suitable for magnetic surveying have been constructed and used in the past. The principal types are described below. It should be noted, however, that two of these, the fluxgate and the proton free-precession magnetometer, have more or less completely replaced the other types in modern work.

2.5.1 Pivoted needle instruments

The sensitive element in these is a magnetic needle arranged to swing on pivots. Probably the oldest known instrument of this type is the *Swedish mine compass* in which a compass needle can rotate in the horizontal as well as the vertical plane and take a position along the total intensity vector. In the *Hotchkiss superdip* the total intensity is determined from the deflection of a system consisting of a magnetic needle to which is fastened a counterarm carrying a small weight and making an adjustable angle with the needle. The system is suspended on a horizontal axle in the magnetic meridian. In the *Thalén–Tiberg magnetometer*, a magnetic needle with an adjustable counterweight is suspended on hardened steel bearings in a small glass-covered case which can be swung on a horizontal axle and held in the vertical or the horizontal plane. The case is first clamped in the horizontal position so that the needle swings on a vertical axis. The direction of the local B_h having been noted from the equilibrium position of the free needle, its magnitude is determined by means of an auxiliary magnet of known moment with the Lamont sine method. The procedure in this method consists essentially in swinging the arm on which the magnet is placed (in a horizontal plane) until the needle and the magnet are at right angles to each other. If the needle has then deflected through an angle θ from its free position, $B_h = B_m /\sin \theta$ where B_m is the known field of the magnet at the centre of the needle. The glass case is then held in a plane perpendicular to the magnetic meridian and the inclination of the needle, which is not vertical owing to the counterweight, is used in combination with the calibration constant to calculate B_z.

2.5.2 Schmidt-type variometers

In the *Schmidt variometer* designed to measure variations in B_z, a magnetic system is free to swing on an agate knife edge in a vertical plane like the beam of a weighing machine. Its equilibrium position at the reference station is adjusted (by altering the centre of gravity) to be horizontal and the deflections from this position at other stations are read by means of an auto-collimating telescope. The instrument can be calibrated by placing it in the field of a pair of Helmholtz coils. In order to eliminate the effect of B_h, the magnetic system is always oriented at right angles to the magnetic meridian when taking observations.

2.5.3 Compensation variometers

These instruments are similar to the Schmidt variometers, but instead of measuring the tilt of the magnetic system from the horizontal they measure the force needed to restore it to that position. The magnetic needle generally hangs on thin wires instead of being balanced on knife edges, and the restoring force is obtained by turning or moving compensating magnets or by means of varying the torsion in the suspension wire. Since the deflection moment due to B_h is zero when the needle is horizontal, the azimuthal orientation of compensation-type variometers is not critical. In variometers designed to measure B_h the magnetic system is initially vertical and restored to this position. The plane of rotation in this case is in the magnetic meridian.

2.5.4 Flux-gate instruments

In these instruments use is made of the fact that the magnetic flux density B induced in certain materials depends non-linearly on the magnetizing force H as shown by the curve (a) in Fig. 2. Strictly speaking the $B-H$ relation is represented by the hysteresis curve (b) in Fig. 2 but it will be sufficient for us to consider the curve (a) in Fig. 2. Increasing the magnetizing force beyond a certain value H_s (the saturating magnetizing force) no longer results in an increase in the flux density in such materials. Within the range $|H| \leqslant H_s$ the $B-H$ curve can be quite accurately described by an equation of the type

$$B = \mu_0 H(a - bH^2) \tag{2.13}$$

where a and b are constants which depend on how quickly the material is magnetically saturated. Now, consider a pick-up coil of effective area A wound tightly around, say, a long thin rod of the material. If the magnetic flux in the rod varies as a result of variations in H, the magnitude of the voltage induced in the coil will be given by

$$V = A \frac{\mathrm{d}B}{\mathrm{d}t} \tag{2.14}$$

If $H = H_0 + p \sin \omega t$ where H_0 is a steady ambient magnetic field and $p \sin \omega t$ is an alternating field produced by an alternating current flowing through an exciting coil wound round the rod, we have from (2.13) and (2.14)

$$V = \mu_0 A\{(a - 3bH_0^2 - \tfrac{3}{4}bp^2)p\omega \cos \omega t - 3bH_0 p^2 \omega \sin 2\omega t$$
$$+ \tfrac{3}{4}bp^3 \omega \cos 3\omega t\} \tag{2.15}$$

It will be seen that the output voltage contains the second and third harmonics in addition to the first but that while the first and third harmonics are always present, the second harmonic appears only if $H_0 \neq 0$. Moreover the amplitude of the second harmonic is directly proportional to H_0. By filtering out the second harmonic in the output voltage and measuring its amplitude, the ambient steady flux density $B_0 = \mu_0 H_0$ in the direction of the rod ('the flux-gate element') can be determined. In contrast to most of the instruments in the previous sections, flux-gate instruments can be made to read absolute flux densities in the element and not just relative ones.

The flux-gate instruments used in practice differ considerably

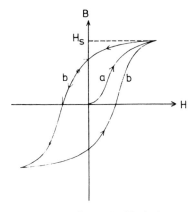

Fig. 2. Non-linear B–H relation.

from each other in construction details but the principle of a very common construction is as follows. The element consists of two juxtaposed, identical cores of a highly permeable magnetic material like mumetal or permalloy, with the windings of the exciting coil in opposite senses around the two cores. A pick-up coil is wound round the entire assembly. Since the signs of the amplitudes p of the two exciting fields are then opposite to each other, the first and third harmonics will be absent in the output voltage from the pick-up coil, as can be seen from Equation (2.15), and a pre-filtered signal is thus obtained. The amplitude of the exciting field is greater than H_s and the frequency is normally a few kHz. Much higher frequencies have, however, been used in some flux-gate instruments. Construction details of some types of flux-gate instruments are available in several papers [3–6].

2.5.5 Proton free-precession magnetometer

Often called 'proton magnetometer' for short, the correct name of this type of instrument is as above and the principle on which it works is as follows. A proton has a magnetic moment **m** as well as an angular momentum **J** (joule second), the relation between the two vectors being

$$\mathbf{m} = \gamma \mathbf{J} \tag{2.16}$$

where $\gamma = 2.67520 \times 10^8$ T^{-1} s^{-1} is an accurately known constant of atomic physics called the gyromagnetic ratio of a proton. The magnetic moment vectors of the protons in, say, a bottle of water in the earth's magnetic field (B) align themselves parallel and anti-parallel to the field. An excess number, in the proportion $\exp(2mB/kT)$ where k is Boltzmann's constant and T is the absolute temperature, will point parallel to the field (the state of lower energy). If a strong magnetic field differing in direction from the earth's field is applied to the water bottle the magnetic vectors gradually align in the direction of the resultant of the two fields. In water it takes approximately five seconds for all the moments to align themselves.

If the additional field is removed rapidly (within about 30 μs), the magnetic moments cannot follow the instantaneous resultant during the removal, and are left in the direction of the original resultant. They are now under the influence of the prevalent earth's field which exerts a torque $mB \sin \theta$ on each proton where θ is the angle between **m** and **B** (Fig. 3). On account of its angular momentum the proton, however, reacts to this torque as would a spinning top, that is, by precessing round the vector **B**. The component of **J** perpendicular to **B** is $J \sin \theta$ and the angular velocity of precession is obtained from the theory of gyroscopes as

$$\omega = \frac{mB \sin \theta}{J \sin \theta} = \gamma B \qquad (2.17)$$

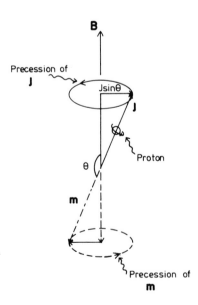

Fig. 3. Precession of protons.

An alternating voltage will be induced in a coil wound round the water bottle, the frequency ν of which can be measured and it follows from (2.17) that

$$B = \frac{\nu}{(\gamma/2\pi)} = 23.4868\nu \ \text{nT} \qquad (2.18)$$

The signal voltage decays exponentially as the precession is damped out and the protons spiral back to their original distribution.

It should be observed that the precession frequency does not depend on the angle between **m** and **B** and hence the direction of the exciting field is immaterial except that it must not coincide with the direction of **B**. If it does no precession will take place and the signal voltage will be zero, but this unlikely situation is easily remedied in practice by simply holding the water bottle in another orientation and taking a new reading. It should also be noted that the quantity measured is *essentially* the frequency of the signal voltage and not its amplitude, so that the orientation of the pick-up coil is also immaterial.

One great advantage of the proton magnetometer is that it requires no levelling. Consequently it is admirably suited to airborne and shipborne measurements. Its disadvantage is that it measures the magnitude of the total ambient field and not its direction. We shall see later that the interpretation of such measurements is inherently more approximate than that of, say, vertical field measurements. For the details of the construction of a proton free-precession magnetometer reference may be made to papers by Waters and Philips [7] and Gupta Sarma and Biswas [8].

2.5.6 High-sensitivity alkali vapour magnetometers

Like the proton magnetometer these instruments, which are also known as optical-pump or absorption-cell magnetometers, exploit the phenomenon of magnetic resonance. Although, for example, helium gas has also been employed in such magnetometers, it is the alkali metal vapours that yield the highest sensitivities and among these rubidium and caesium are the most widely used elements. The atoms of alkali metals possess a magnetic moment mainly due to the spin of the single periferal electron but also due to the spin of the nucleus. Each atom is normally in a ground state of energy and there are also (discrete) energy states to which the atom may be excited. When the atom is placed in a magnetic field each energy state is split into several discrete sub-levels (the so-called Zeeman effect) corresponding to the possible discrete orientations that the magnetic moment may take with respect to the magnetic field **B**. The energy difference between two split levels belonging to any state (the ground state as well as the possible higher states) is proportional to B. A very simplified scheme of the Zeeman effect in an alkali atom is shown in Fig. 4a where the numbers $-\frac{1}{2}$, $\frac{1}{2}$ . . . etc. represent the spin quantum numbers m of the atom. They may be looked upon conveniently as proportional to the projections of the atomic magnetic moment, in one of its possible orientations, on the magnetic field direction.

If now the atom is illuminated with right-handed circularly polarized light of proper frequency an atom in a ground state is excited to a higher state, but according to the quantum theory only transitions for which $\Delta m = +1$, e.g. those from $m = -\frac{1}{2}$ to $m = +\frac{1}{2}$, take place. From this excited state the atom will (after about 10 ns) emit radiation in returning to the ground state. In this transition, however, the atom may return to a ground state level with $m = +\frac{1}{2}$ as well as $m = -\frac{1}{2}$, but the transition probability to $m = +\frac{1}{2}$ is greater. Moreover, a spontaneous transition from $m = +\frac{1}{2}$ to $m = -\frac{1}{2}$ in the ground state levels is forbidden by the quantum theory. The result is that if a cell of alkali atoms in a

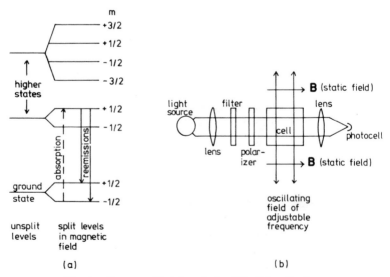

Fig. 4a. Zeeman effect (much simplified) in a Cs atom.

Fig. 4b. Scheme of an alkali vapour magnetometer.

magnetic field is illuminated continuously with circularly polarized light of appropriate frequency, the ground state levels with
$m = +\frac{1}{2}$ will be gradually filled at the expense of levels having
$m = -\frac{1}{2}$ which will be ultimately emptied. This process, discovered
by the French physicist A. Kastler in 1950, is called *optical
pumping*. Evidently since no transitions to the excited level can
take place after the ground state level with $m = -\frac{1}{2}$ is empty, no
energy will be absorbed from the incident light, the 'pump' will
stop and the intensity of the light transmitted through the cell will
increase to a maximum.

Suppose now that an electromagnetic field of adjustable
frequency is applied to the cell in a direction different from that
of **B**. If the frequency ν is such that $h\nu = \Delta E$ where h is Planck's
constant and ΔE is the energy difference between the ground state
levels $m = +\frac{1}{2}$ and $m = -\frac{1}{2}$, the atoms absorb energy from the field
and transitions take place to $m = -\frac{1}{2}$, the pump restarts and the
intensity of light transmitted through the cell will drop sharply.
However, ν can be shown to be equal to the precession frequency
of the atomic moment m_a and the atomic spin J_a (Appendix 2)
around B, and in analogy with the proton precession of the
previous section we get

$$2\pi\nu = \frac{m_a B}{J_a}$$

or

$$B = \frac{\nu}{2\pi(m_a/J_a)} \tag{2.19}$$

The factor $2\pi(m_a/J_a)$ is known accurately for many atoms. For
the naturally occurring isotopes $^{37}_{85}$Rb and $^{55}_{133}$Cs the values are
4.67 and 3.498 Hz/nT. Like proton magnetometers the vapour
magnetometers measure only the magnitude of the ambient field.
Also they do not need any levelling. Since ν can be determined
very accurately on account of the sharpness of the resonance,
sensitivities of the order of 0.01 nT can be obtained with alkali
vapour magnetometers. The construction of a caesium magnetometer has been described in some detail by Giret and Malnar [9].
They used an absorption cell of 100 ml volume containing free
atoms of Cs in equilibrium with the metal, at a pressure of about
267 μPa (2 nmHg). Fig. 4b shows the general scheme of an alkali
vapour magnetometer.

2.5.7 Some general remarks

The accuracy of pivoted needle instruments is scarcely better than ±100 nT and they are now seldom used. The Schmidt- and compensation-type variometers are precision instruments having accuracies better than ±5 nT, although the accuracy of some instruments of this type, intended for rough reconnaissance surveys, is not better than ±20–50 nT. The flux-gate, proton free-precession and alkali vapour magnetometers have the advantage that their sensitive elements and the measuring or the recording system can be widely separated by cables. They can, therefore, be conveniently used under certain circumstances (e.g. in boreholes, under water, in airborne work, etc.) where the design of older instruments would be highly complicated. Their other characteristic is the speed of measurement, some 10 s, in contrast to the older instruments which may need a minute or so for one measurement.

The accuracy of flux-gate and proton magnetometers is of the order of 1 nT although the requirement of levelling for flux-gate instruments leads in practice to accuracies more like ±5–10 nT and much less in borehole and airborne work. Proton magnetometers of special construction give accuracies of the order of 0.1 nT, while alkali vapour instruments are superior in accuracy by a factor of 10 as has already been seen. Flux-gate instruments have the advantage over both these that they can measure *components* of the flux density. It is also worth noting that if the external magnetic field is strongly inhomogeneous the decay of proton precession is so fast that no signal is obtained. The limit of admissible gradient in the best designs appears to be about 500 nT per m. The high sensitivity of alkali vapour magnetometers can be useful for surveys in areas (e.g. sedimentary basins) where magnetic field variations are small from point to point, provided uncertainties in measurements due to other causes can be avoided.

2.6 Relative merits of ΔB_h, ΔB_z and ΔB_t measurements

Among the magnetic elements the direction of the field is the element least sensitive to changes in the dimensions and magnetic properties of a sub-surface body. It is therefore never used by itself in accurate work. Of the remaining, namely B_h, B_z and B_t, any one or more can be chosen for measuring the respective anomalies ΔB_h, ΔB_z and ΔB_t. Since ΔB_h and ΔB_t are associated with a change in direction as well, the interpretation of these anomalies becomes somewhat complicated. The anomalous vector in the total field is $\mathbf{B}_t - \mathbf{B}_{0t}$ where $\mathbf{B}_{0t}$ is the normal flux density in the area and the magnitude of the vector is

$$| \mathbf{B}_t - \mathbf{B}_{0t} | = (B_z^2 + B_h^2)^{1/2} \tag{2.20}$$

However, since proton or alkali vapour magnetometers measure only the magnitudes B_t and B_{0t}, the anomaly obtained from measurements with them is not (2.20) but

$$\Delta B_t = B_t - B_{0t} = (B_z^2 + B_h^2)^{1/2} - (B_{0z}^2 + B_{0h}^2)^{1/2} \tag{2.21}$$

If α is the declination of $\Delta \mathbf{B}_h$ (Fig. 5),

$$B_h^2 = B_{0h}^2 + 2B_{0h} \Delta B_h \cos \alpha + \Delta B_h^2$$

Also,

$$B_z^2 = (B_{0z} + \Delta B_z)^2 = B_{0z}^2 + 2B_{0z} \Delta B_z + \Delta B_z^2$$

Hence

$$\Delta B_t = (B_{0t}^2 + 2B_{0h} \Delta B_h \cos \alpha + 2B_{0z} \Delta B_z + \Delta B_t^2)^{1/2} - B_{0t} \tag{2.22}$$

We can simplify (2.22) to

$$\Delta B_t = \frac{B_{0h}}{B_{0t}} \Delta B_h \cos \alpha + \frac{B_{0z}}{B_{0t}} \Delta B_z$$

$$= \Delta B_h \cos \alpha \cos I + \Delta B_z \sin I \tag{2.23}$$

where I is the normal geomagnetic inclination in the area.

From the theoretical standpoint measurements of ΔB_z are to be preferred to those of ΔB_h or ΔB_t since the sub-surface picture can be visualized much more readily from ΔB_z but the present

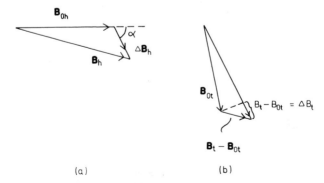

(a)   (b)

Fig. 5. Normal and anomalous vectors of total flux density.

tendency in geophysical surveys, even on ground, is strongly towards using proton magnetometers for determining ΔB_t, owing no doubt to the great convenience that levelling or orientation is not necessary and the speed of surveys is higher than for ΔB_z measurements of comparable accuracy. Sometimes both ΔB_z and ΔB_h are measured (for example, by suitable flux-gate instruments). In a survey over an area which may be considered to be virtually an infinite plane, such a procedure offers no additional information if observations of only ΔB_z are made sufficiently densely, and is therefore superfluous because both ΔB_h and ΔB_t can then be calculated. This rather surprising result, the converse of which is not true (except for the case when ΔB_h is determined), follows from Green's theorem in potential theory and will be proved in the chapter on gravitational methods. However, in certain special cases (e.g. in underground work) where the measurements cannot be carried out over a plane (or a closed surface) or where a sufficiently dense network of points is not available, the knowledge of both ΔB_h and ΔB_z can be of some additional help.

Another surprising result of practical importance is that the total change in ΔB_z across geological features such as thin dikes or spherically shaped masses is always (that is, even in low magnetic latitudes) greater than or equal to that in ΔB_h (cf. Figs. 8 and 12). Thus, from this point of view too, measurements of ΔB_z are to be preferred to those of ΔB_h *all over the world*.

2.7 Field procedure

When the area for magnetic investigations has been selected a base line is staked parallel to the geological strike (the trace of the rock strata on the plane represented by the surface of the earth) and measurements are made at regular intervals along lines perpendicular to the base. A reference point, far from artificial disturbances such as those due to power lines, railways, etc. is chosen and the magnetic intensity values at all other points are measured as positive or negative differences from the intensity at this point. The reference point may, strictly speaking, be anywhere within or without the area, but it is convenient to choose a point, as near as possible to the area or inside it, at which the magnetic field is known to be approximately the normal field.

Certain precautions must be taken in magnetic measurements. Magnetic materials in the wearing apparel of the observer, like keys, penknives, wrist-watches (also 'non-magnetic' ones!), etc.

can totally vitiate the observations and ought to be entirely removed. Observations in the neighbourhood of iron objects such as motor cars, iron-junk, railroads, bridges should be avoided.

Several corrections must generally be applied to a set of magnetometer observations as follows:

(1) To correct for the diurnal variations of the earth's field, an auxiliary instrument is kept at some convenient station within the area and read at intervals. Automatic recording may also be used. The algebraic difference between the readings of this instrument at any time, t, and the reference time, $t = 0$ (corrected, if necessary for the temperature coefficient) is subtracted from the reading at the field station, which was also measured at time t. If great accuracy is not desired, it is sufficient to repeat at $1-2$ hours' interval the reading at a previously occupied station and distribute the difference over the stations measured during this interval. The correction thus obtained includes both the diurnal variation and the temperature coefficient of the instrument.

(2) Corrections for the temperature dependence of the field instrument can be applied if its temperature coefficient is known. Most modern instruments are, however, temperature compensated and the correction is therefore negligible.

(3) Terrain corrections: Rough terrain may give rise to anomalies. For instance, if the rocks above a station situated in a depression are magnetic, 'false' negative anomalies will be recorded. There are no general rules for applying terrain corrections. Usually, the anomalies showing a strong correlation with the terrain are regarded as less significant than others.

An interesting approach to reducing the measured magnetic anomalies to a single horizontal plane, above or through the highest point of the topography, to overcome topographic effects has been indicated by Roy [10]. But it requires a Fourier transformation of the field and is probably not practical to use for routine processing.

An important point in considering the anomalies in an area is the zero level, that is the readings of the instrument at points where the field is the normal undisturbed geomagnetic field. If the readings remain constant, or vary randomly so as to suggest a geologic noise only, over a sufficient length of a measuring profile, say $100-500$ m, the reading at any point on this stretch may be taken to be the zero reading and the anomalies at all other points in the area referred to it. If distinct magnetically anomalous masses are evident in the area, the zero level can be determined from the flanks of an anomaly curve, since they approach it asymptotically at great distances from the mass. Regional anomaly gradients, terrain characteristics, or contacts between rock formations of differing magnetizations, sometimes make it impossible to use the same zero level throughout the area.

2.8 The interpretation of magnetic anomalies

The magnetic anomaly field A satisfies Laplace's equation in potential theory, namely

$$\nabla^2 A = 0 \qquad (2.24)$$

This equation does not suffice to find the sub-surface magnetization uniquely (see further, Chapter 3). The usual procedure in interpreting magnetic (or gravity) anomalies is to guess a body of suitable form, calculate its field on the surface and compare it with observations. It is then possible to adjust the depth and dimensional parameters of the body by trial and error until a satisfactory fit is obtained. Such a solution is only one of an infinity of possible solutions. The assumption very often made in magnetic calculations is that of homogeneous magnetization which is, however, only valid for bodies bounded by second degree surface (spheres, ellipsoids, infinitely long cylinders, etc.). Geologic structures do not conform to these shapes except in very rare instances and even then only approximately. At points inside a geologic body far from its bounding surface, the above assumption may be more or less true, but at the edges and corners it definitely fails. Consequently, all calculations of the magnetic field of a geologic structure are approximations of varying degree which conform better to the true field the greater the distance from the structure.

A first step towards interpretation is the preparation of a 'magnetic map' on which the intensity values at different stations are plotted and on which the contours of equal ΔB_h, ΔB_z or ΔB_t (isoanomalies) are drawn at suitable intervals. The interpolation necessary in contour drawing is particularly easy if the observation points form a square network. In such cases, the most accurate contours are obtained if the interpolation is carried out mainly along the approximate direction of the isoanomalies, since this is the direction in which the gradients are low. A trial isoanomalous line is first sketched to obtain the trend and is subsequently corrected by more exact interpolation. Contouring of geophysical maps, especially of large scale surveys like airborne surveys, is nowadays often done on automatic plotters using computer programs for interpolation.

Certain qualitative conclusions are readily drawn from a magnetic map. Anomalous conditions in the sub-surface are indicated, for instance, by successive closed contours with the anomaly-values increasing or decreasing towards a 'centre', while the direction of elongation of the closed curves may be identified with the strike of the anomalous body. Another indication is given by high horizontal anomaly gradients. They are often associated with contacts between rocks of different susceptibilities or unequal total intensities of magnetization, the contact lying shallower the steeper the gradient.

As regards strike it should be noted that in low magnetic latitudes the above rule of thumb should be used with great caution, because in these regions, bodies of finite length, striking north–south, produce anomaly patterns which indicate an apparent east–west strike. The effect is easily understood by considering the field created by the longitudinal magnetization of the body by the horizontal component of the earth's magnetizing force. In fact, the anomaly pattern in this case closely resembles that in Fig. 7 discussed later. For strike directions of the body deviating from the north too, the anomaly patterns in low magnetic latitudes indicate apparent strikes significantly different from true ones [11]. In high magnetic latitudes the effect of a strong, almost horizontal remanent magnetization may similarly cause the anomaly patterns to deviate from the true strike.

Despite the non-uniqueness of the solutions of Equation (2.24) the interpretations of magnetic anomalies is not so seriously hampered in practice as might be imagined. The reason is that on the basis of geological information, which is usually available, or on grounds of plausibility, it is normally possible to reduce the number of alternatives to a moderate one from which only a few need be selected initially as working hypotheses for the trial calculations.

2.8.1 Anomalies of spheres

Starting from Equation (2.6) it can be shown (Appendix 3) that a sphere can be magnetized homogeneously by a uniform magnetizing force (such as the earth's) in the direction of the force and that its potential is the same as that of a dipole of identical moment placed at the centre of the sphere. We shall consider the general case of a sphere whose magnetic moment vector **m** is not necessarily in the direction of T_0, the earth's undisturbed magnetizing force, thus implying a remanent magnetization besides that induced by T_0.

Let i ($\neq I$) be the inclination of **m** with the horizontal and ϵ the angle between the vertical plane through **m** and the magnetic meridian. Other notations will be (cf. Fig. 6):

x = coordinate along the earth's surface measured from an origin O vertically above the sphere's centre,

a = depth of centre below the surface,

$r^2 = a^2 + x^2$ (square of distance from centre) and

γ = angle between the magnetic meridian and the line Ox.

The z-axis is directed downwards. The moment **m** can be resolved into a vertical component of magnitude $m \sin i$ and two horizontal components of magnitude $m \cos i \cos(\gamma - \epsilon)$ and $m \cos i \sin(\gamma - \epsilon)$. With the angles as shown in Fig. 6, the potential at x can be written down from Equation (2.8) as

$$\phi = \frac{m \sin i \cos \theta}{4\pi} \frac{\cos \theta}{r^2} + \frac{m \cos i \cos (\gamma - \epsilon)}{4\pi} \frac{\cos \psi}{r^2}$$

$$= -\frac{m \sin i}{4\pi} \frac{\sin \psi}{r^2} + \frac{m \cos i \cos (\gamma - \epsilon)}{4\pi} \frac{\cos \psi}{r^2} \qquad (2.25)$$

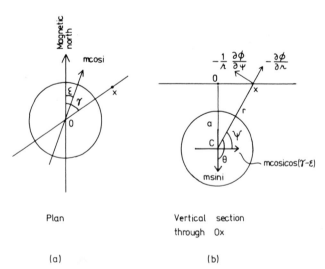

Plan   Vertical section through Ox

(a)   (b)

Fig. 6. Magnetized sphere.

The component $m \cos i \sin(\gamma - \epsilon)$ does not contribute to ϕ as it makes an angle of $\pi/2$ with Cx. It does, however, contribute to the anomalous flux density at x by the amount

$$\Delta B_y = \mu_0 \frac{m \cos i \sin (\gamma - \epsilon)}{4\pi r^3} \quad (2.26)$$

directed outwards from the plane of Fig. 6b, that is, in the y-direction (cf. Appendix 3).

From the components of the magnetizing force $-(\partial\phi/\partial r)$ and $-(1/r)(\partial\phi/\partial\psi)$ along and perpendicular to Cx we get the anomalous flux densities due to the sphere

$$\Delta B_x = -\mu_0 \frac{\partial\phi}{\partial r} \cos \psi + \mu_0 \frac{1}{r} \frac{\partial\phi}{\partial\psi} \sin \psi$$

$$\Delta B_z = \mu_0 \frac{\partial\phi}{\partial r} \sin \psi + \mu_0 \frac{1}{r} \frac{\partial\phi}{\partial\psi} \cos \psi \quad (2.27)$$

Using (2.25) we easily obtain

$$\Delta B_x = \frac{\mu_0}{4\pi} \frac{m \cos i}{r^3} \left[\left(\frac{3x^2}{r^2} - 1 \right) \cos (\gamma - \epsilon) \right.$$

$$\left. - \frac{3ax}{r^2} \tan i \right] \quad (2.28)$$

$$\Delta B_z = \frac{\mu_0}{4\pi} \frac{m \sin i}{r^3} \left[\frac{3a^2}{r^2} - \frac{3ax}{r^2} \cot i \cos(\gamma - \epsilon) - 1 \right] \quad (2.29)$$

To calculate ΔB_t in the sense of Equation (2.23) we must, of course, first calculate $\Delta B_h = (\Delta B_x^2 + \Delta B_y^2)^{1/2}$.

If there is no remanent magnetization, $\mathbf{m}$ lies in the magnetic meridian ($\epsilon = 0$, $i = I$) and its magnitude is (cf. Appendix 3)

$$\frac{4}{3} \pi b^3 \frac{3\kappa}{3 + \kappa} T_0 \quad (\text{A m}^2)$$

where b is the radius of the sphere and κ the susceptibility. In this case, for a profile along the magnetic meridian ($\gamma = 0°$) we get from (2.28), (2.29) and (2.23):

$$\Delta B_t = \frac{\mu_0}{4\pi} \frac{m}{r^3} \left[\frac{3x^2 \cos^2 I - 3ax \sin^2 I + 3a^2 \sin^2 I}{r^2} - 1 \right] \quad (2.30)$$

The lines of equal ΔB_h and ΔB_z due to a sphere just touching the surface of the earth ($a = b$) at a place where B_{0z} = 45 000 nT and B_{0h} = 20 000 nT ($I = 66°$) are shown in Fig. 7.

In Fig. 8a are reproduced the profiles of ΔB_z and ΔB_h along the lines of measurement passing through the origin and lying in the magnetic meridian over the spheres with different directions of magnetization. The abscissae are in units of a. It will be seen that north of the magnetic equator ($I > 0$), the maximum positive ΔB_z as well as ΔB_h occurs south of the point vertically above the centre of the sphere. The conditions south of the magnetic equator are the reverse. In both cases, the displacement of the absolute maximum anomaly in ΔB_z and ΔB_h from the origin is greater, the smaller the inclination. Fig. 8b shows the corresponding profiles for ΔB_t calculated from Equation (2.30). In general a ΔB_h, ΔB_z or ΔB_t profile possesses three extrema here. If s_1, s_2, s_3

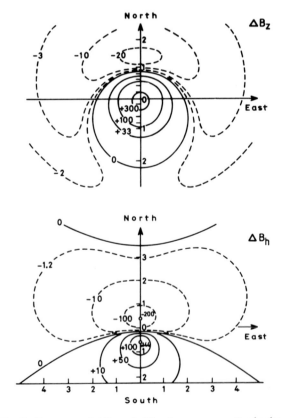

Fig. 7. Contours of ΔB_z and ΔB_h above a magnetized sphere.

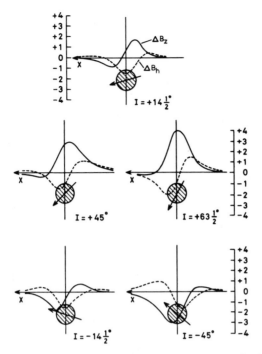

Fig. 8a. Profiles of ΔB_z and ΔB_h above a magnetized sphere.

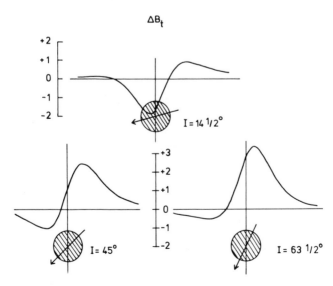

Fig. 8b. Profiles of ΔB_t above a magnetized sphere.

are the three mutual distances between the points at which the extrema occur we obtain the following exact expressions for the sum of the squares $s_1^2 + s_2^2 + s_3^2$.

$$\Delta B_h : \Sigma s^2 = 9\{4 \tan^2 i \sec^2 (\gamma - \epsilon) + 1\}a^2 \qquad (2.31a)$$

$$\Delta B_z : \Sigma s^2 = 8\{4 \cot^2 i \cos^2 (\gamma - \epsilon) + 3\}a^2 \qquad (2.31b)$$

The corresponding result for ΔB_t for the case represented by (2.30) is

$$\Sigma s^2 = \frac{126 \sin^4 I - 106 \sin^2 I + 36}{(1 - 3 \cos^2 I)^2} a^2 \qquad (2.31c)$$

These expressions follow from the cubic equations in x obtained by setting the first derivatives of ΔB_h, ΔB_z and ΔB_t with respect to x equal to zero, and using the following well-known result in the theory of equations. If x_1, x_2, x_3 are the roots of $ax^3 + bx^2 + cx + d = 0$, then;

$$(x_1 - x_2)^2 + (x_2 - x_3)^2 + (x_3 - x_1)^2 = 2\frac{b^2}{a^2} - 6\frac{c}{a} \qquad (2.32)$$

Since the algebra is straightforward the details may be left to the interested reader.

Equations (2.31) furnish simple formulae for estimating the depth a if the three extrema can be located sufficiently accurately. Other rules can be formulated from the distance at which the anomaly or its slope falls to a certain fraction of its maximum value etc. but they tend to be algebraically much more involved that Equations (2.31). Exact expressions similar to (2.28) and (2.29) may also be obtained for long, horizontal cylinders and ellipsoids [12, 13]. For ellipsoids, however, they are very complicated. But in both these cases the anomalies are similar in their general features to those of a sphere. Of course, geologic bodies cannot be expected to assume these simple shapes, but curves such as those in Figs. 7 and 8 are of considerable help in qualitative interpretation.

2.9 Geological features

Some further shapes by which geological features can be approximated are shown in Fig. 9. The step structure in (a), assumed to be very long perpendicular to the plane of the paper, is often encountered as a fault when rock beds have been thrust up or

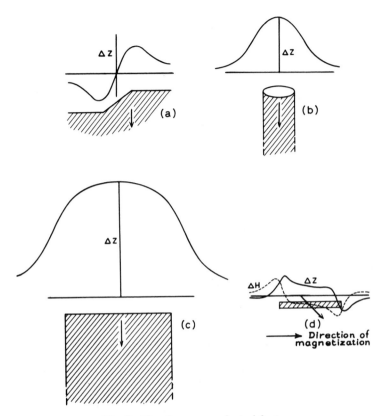

Fig. 9. Miscellaneous geological features.

down with respect to each other along a slip-plane. It may also represent the undulations of the bedrock under an overburden, a flexure of rock beds or an erosional hill—valley combination. The cylinder in (b) may be a plug of salt or intrusive volcanic rock. The thick plate in (c) may approximate to a broad zone ('sheet') of magnetically impregnated rock or a thick dike, while the plate in (d) may be encountered as a horizontally-lying feature. Typical magnetic anomalies over them are also qualitatively shown.

Of the various features shown, (a) and (c) merit further attention on account of their widespread occurrence geologically. We shall deal with a more general case than (c) in which the sides are not vertical, but sloping. In treating its magnetic anomaly it is, however, convenient to start with another feature, namely, a thin sheet which is often a good approximation to many ore veins and dikes.

2.10 Anomalies of sheets and prisms

2.10.1 Thin sheets

Figs. 10a and b show a uniformly magnetized thin dipping sheet of infinite length and depth extent and of width b in plan and section. The magnetization vector **M** makes an angle i with the horizontal, while the vertical plane through it makes an angle ϵ with the magnetic meridian and δ with the strike direction of the sheet. Such a sheet will have a vertical magnetization of $M \sin i$, a horizontal magnetization in the strike direction of $M \cos i \cos \delta$ and a horizontal magnetization of $M \cos i \sin \delta$ perpendicular to both of these.

It will be seen from Fig. 10b that the component of M in the cross-section of the sheet is $M' = M(1 - \cos^2 i \cos^2 \delta)^{1/2}$ inclined at an angle $i' = \tan^{-1}(\tan i / \sin \delta)$. Choosing x- and z-axes perpendicular and parallel to the sheet as shown, the potential at $P(\xi, \zeta)$ can be found by integration of the potential $d\phi$ due to an element of the sheet having the infinitesimal cross-section $b\, dz$. If x, z are the coordinates of such an element then (Appendix 4),

$$d\phi = -\frac{1}{4\pi} M_x b\, dz\, \frac{\partial}{\partial x} \ln \rho^2 - \frac{1}{4\pi} M_z b\, dz\, \frac{\partial}{\partial z} \ln \rho^2$$

where

$$\rho^2 = (\xi - x)^2 + (\zeta - z)^2$$

If the origin is chosen at the top edge of the sheet, we can put $x = 0$ after differentiation, obtaining

$$d\phi = \frac{1}{4\pi} M_x\, 2b\, dz\, \frac{\xi}{\xi^2 + (\zeta - z)^2} + \frac{1}{4\pi} M_z b\, dz\, \frac{2(\zeta - z)}{\xi^2 + (\zeta - z)^2}$$

Integration between $z = 0$ and $z = \infty$ gives

$$\phi = \frac{1}{4\pi} 2M'_{\perp} b \left(\frac{\pi}{2} + \tan^{-1} \frac{\zeta}{\xi} \right) + \frac{1}{4\pi} M'_{\parallel} b \ln(\xi^2 + \zeta^2) + C_\infty \quad (2.33)$$

where C_∞ is an infinite constant and the notation M_x, M_z is replaced by $M'_{\perp}$ and $M'_{\parallel}$ to denote magnetizations (components of M') perpendicular and parallel to the sheet.

The magnetizing force in any direction whose direction cosines are l, n with respect to the ξ and ζ directions is

$$\left(-\frac{\partial \phi}{\partial \xi} \right) l + \left(-\frac{\partial \phi}{\partial \zeta} \right) n \quad (2.33a)$$

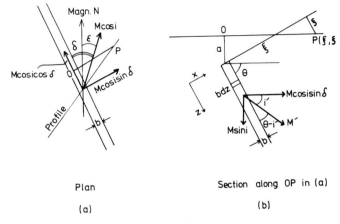

Plan

Section along OP in (a)

(a)

(b)

Fig. 10 (a and b). Thin, magnetized sheet in plan and section. (Note that in the final formulae x is used to denote the distance OP and not a coordinate with respect to the x-, z-axes shown).

In particular, $(l, n) \equiv (\sin \theta, \cos \theta)$ for the horizontal direction and $(-\cos \theta, \sin \theta)$ for the vertical direction. It is then a matter of elementary algebra to show that the flux densities in the two directions are

$$\Delta B_{h} = \frac{\mu_0}{4\pi} 2b \frac{xM'_{\parallel} + aM'_{\perp}}{a^2 + x^2} \qquad (2.34)$$

$$\Delta B_{z} = \frac{\mu_0}{4\pi} 2b \frac{aM'_{\parallel} - xM'_{\perp}}{a^2 + x^2} \qquad (2.35)$$

where x is now used to denote the distance of P from the point O (Fig. 10b) on the same level as P and is positive in the direction from which θ is measured. If there is no remanence, $\epsilon = 0$, $i = I$ and $i = \tan^{-1} (\tan I/\sin \delta) = I'$(say). In this case the azimuth of ΔB_h is $(90° - \delta)$ since, by symmetry, there cannot be a horizontal flux density in the strike direction so that (cf. (2.23)),

$$\Delta B_{t} = \Delta B_{h} \sin \delta \cos I + \Delta B_{z} \sin I \qquad (2.36)$$

Since in this case $M' = \kappa T'_0$ where T'_0 is the component of the total magnetizing force of the earth's field in the cross-section of the sheet

$$M'_{\perp} = M' \sin(\theta - I')$$
$$M'_{\parallel} = M' \cos(\theta - I')$$

Equations (2.34) and (2.35) can then be written as

$$\Delta B_h = \frac{b \kappa B_0' \cos(\theta - I')}{2\pi} \frac{x + a \tan(\theta - I')}{a^2 + x^2} \tag{2.34a}$$

$$\Delta B_z = \frac{b \kappa B_0' \cos(\theta - I')}{2\pi} \frac{a - x \tan(\theta - I')}{a^2 + x^2} \tag{2.35a}$$

where $B_0' = B_0(1 - \cos^2 I \cos^2 \delta)^{1/2}$ is the component of the earth's normal flux density B_0 in the cross-section of the sheet. If the magnetization is almost perpendicular to the sheet, as may happen in low magnetic latitudes for steeply dipping sheets, it is preferable to rewrite Equations (2.34a) and (2.35a) with $\sin(\theta - I')$ instead of $\cos(\theta - I')$ as a factor. ΔB_t may be similarly expressed using (2.36).

The magnetic anomaly profiles at right angles to the strike of a long thin sheet are similar in general shape to those shown in Fig. 8 except that for the sheet there are only two extreme points (one maximum and one minimum), three inflection points and one zero-crossing ($\Delta B_z = 0$). The position of the top of the sheet, the depth a and the parameter $M_\perp'/M_\parallel$ ('cross-magnetization' ratio) can be determined from the relative positions of these characteristic points.

For a thin sheet for which $M_\perp' = 0$ and in which there is no remanent magnetization, (2.35a) becomes

$$\Delta B_z = \frac{b \kappa B_0'}{2\pi} \frac{1}{1 + x_a^2} \tag{2.35b}$$

where $x_a = x/a$. It is obvious that in this case

$$\Delta B_z(\text{max}) = \frac{b \kappa B_0'}{2\pi a} \tag{2.35c}$$

$$a = x_{1/2}$$

where $x_{1/2}$ is the distance at which $\Delta B_z = \Delta B_z(\text{max})/2$. The depth and the parameters $b\kappa$ may be calculated from (2.35c).

Equations corresponding to (2.35b) can be derived also for the general case of a thin plate of strike length L. The ΔB_z-anomaly along an arbitrary orthogonal profile across a plate of finite length can be shown to be (exactly)

$$\Delta B_z(L) = \Delta B_z(\infty)[f(d_{1a}) + f(d_{2a})] \tag{2.35d}$$

where $d_{1a} = d_1/a$ and $d_{2a} = d_2/a$ are the distances of the profile from the two ends of the plate and

$$f(d_a) = \frac{d_a}{(x_a^2 + d_a^2 + 1)^{1/2}} \tag{2.35e}$$

$x_{1/2}$ referred to below is now defined for the central profile.

If we define a distance $y_{1/2}$ (analogous to $x_{1/2}$) along a line parallel to and directly above such a plate, the ratio $x_{1/2}:y_{1/2}$ depends on L/a and has the limiting values 1.000 ($L = 0$) and 0 ($L = \infty$). Similarly $a/x_{1/2}$ will be a funtion of L/a with the limiting values 1.305 and 1.000 respectively. In Fig. 10c are plotted $x_{1/2}:y_{1/2}$, $a/x_{1/2}$ and the quantity $b\kappa B_{0z}/(4\pi x_{1/2}\Delta B_z(\text{max}))$ (equal from (2.35c) to 1/2 for an infinite thin plate), against L/a. If $x_{1/2}:y_{1/2}$ can be determined from the observed anomaly profiles, the depth to the upper surface, length and 'magnetic width' ($b\kappa$) of a body which can be approximated by a thin vertical or steeply dipping plate of very large depth extent and no transverse magnetization, may be estimated from Fig. 10c.

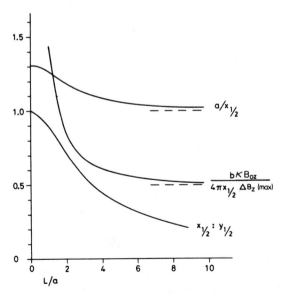

Fig. 10c. Parameters of a thin sheet having no transverse magnetization, as a function of sheet length.

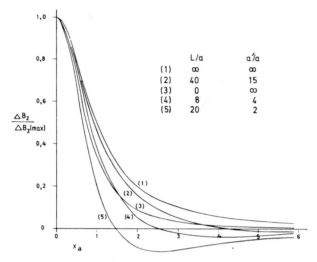

Fig. 10d. ΔB_z across some thin vertical sheets of various depth extents and lengths.

If it is desired to take into account the effect of a finite depth extent, ΔB_z may be calculated from Equations (2.35b) and (2.35d) by subtracting from their right-hand sides expressions of identical forms but with a', the depth of the lower surface, replacing a. Some examples of anomaly curves of such magnetic plates are shown in Fig. 10d. The positions of the zero and the negative minimum through which ΔB_z passes in these cases depends on L/a and a'/a and can be used to estimate the depth extent. The ratio $a/x_{1/2}$ depends likewise on L/a and a'/a (Table 5) so that the depth of the upper surface cannot be calculated without a reasonable assumption about the depth of the lower one. The effect of an appreciable deviation of the dip from

Table 5 $a/x_{1/2}$ for magnetic double lines

L/a	a'/a			
	1	2	4	∞
0	1.99†	1.54	1.37	1.305
2	1.91‡	1.45	1.26	1.18
4	1.88‡	1.43	1.21	1.08
8	2.02‡	1.48	1.20	1.03
∞	2.06‡	1.53	1.26	1.00

†Point dipole. ‡Linear dipoles.

the vertical may be taken into account by shifting the lower surface sideways. In that case the negative minima in Fig. 10d will be greater in magnitude on the dip side than on the opposite one.

2.10.2 Sheet of arbitrary thickness

A long sheet of arbitrary thickness and infinite depth extent may be considered to be built up of an infinite number of thin sheets and the anomaly found by integration. Starting from Equation (2.33) a general result can be obtained for the anomaly $\Delta B(l, 0, n)$ in any arbitrary direction whose direction cosines are $(l, 0, n)$ at P (Fig. 11, where the y-axis is perpendicular to the plane of the paper and the z-axis is vertically downwards). We have

$$\Delta B(l, 0, n) = \left(\frac{\mu_0}{4\pi}\right) 2M' \sin\theta \left[\{n\cos(\theta - i') - l\sin(\theta - i')\}\right.$$
$$\times (\alpha_1 - \alpha_2) - \{l\cos(\theta - i') + n\sin(\theta - i')\}$$
$$\left.\ln(r_1/r_2)\right] \tag{2.37}$$

M' and i' have the same significance as in the previous section and x is positive in the direction from which θ is measured $(0° < \theta < 180°)$. For angles of dip within about $20°$ from the horizontal, however, the geometry of a thick sheet as in Fig. 11 tends to become geologically implausible. The derivation of Equation (2.37) is given in Appendix 5. If in (2.37) $l = 1$, $n = 0$ then $\Delta B = \Delta B_h$, while if $l = 0$, $n = 1$ then $\Delta B = \Delta B_z$. Equation (2.37) can be written as

$$\Delta B(l, 0, n) = \left(\frac{\mu_0}{4\pi}\right) 2M'C \sin\theta \{(\alpha_1 - \alpha_2) - k\ln(r_1/r_2)\} \tag{2.38}$$

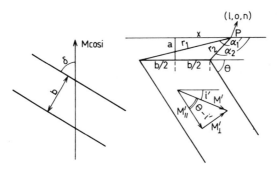

Fig. 11. Sheet of arbitrary thickness in plan and section.

where $2(\mu_0/4\pi)M'C\sin\theta$ may be called an 'amplitude factor' and k a 'shape factor'. The values of C and k for ΔB_h, ΔB_z and ΔB_t are as follows:

Field	C	k
ΔB_h	$-\sin(\theta - i')$	$-\cot(\theta - i')$
ΔB_z	$\cos(\theta - i')$	$\tan(\theta - i')$
ΔB_t	$\dfrac{\sin I}{\sin I'}\sin(I' + i' - \theta)$	$\cot(I' + i' - \theta)$

Figs. 12a and b show ΔB_h and ΔB_z curves across a sheet ($b = 2a$) dipping at various angles. These are quite general as far as strike angle δ and magnetization direction are concerned. If the direction of remanence coincides with the earth's field, $i = I$ and $i' = I'$ so that these curves are then valid for strike angles $\delta = \sin^{-1}$ (tan $I/$ tan I') measured from the magnetic meridian (positive anti-clockwise in plan). The curves labelled $\Delta B_t'$ in Figs. 12a and b show the total field anomaly for east—west strike and no remanence.

It will be seen that, except when $\theta - i'$ takes certain values, each curve possesses two extrema (mutual distance $= E$). By expressing the α's (as arctangents) and r's in (2.38) in terms of x, a and b it can be shown by straightforward differentiation that E, a and b are related by the equation

$$a = \frac{\sqrt{(E^2 - b^2)}}{2} \frac{|k|}{\sqrt{(1 + k^2)}} \tag{2.39}$$

where k is the appropriate parameter in the table above. Obviously a, b and k cannot be separately determined from a knowledge of E alone. Rules for estimating a based on various characteristic abscissae of the anomaly curves have been proposed by various workers and reviewed by Åm [14].

Hood [15] has, among others, also dealt with the interpretation of the magnetic anomalies of the dipping sheet. Although his

---

Fig. 12a. Magnetic anomalies ΔB_z, ΔB_h and ΔB_t across a thick sheet ($b = 2a$). In the absence of remanence or with remanence coinciding with the direction of the earth's field, $I' = \tan^{-1}$ (tan $I/\sin\delta$) where δ is the strike angle. All ΔB_t curves shown are, however, for E—W strike of sheet only ($\delta = 90°$, $I' = I$) and no remanence.

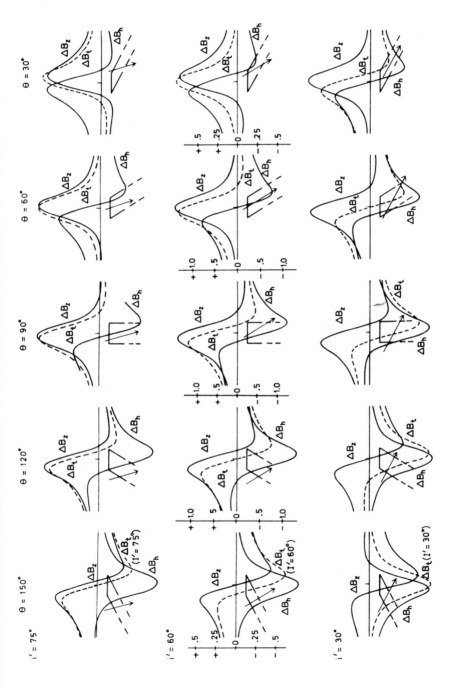

43

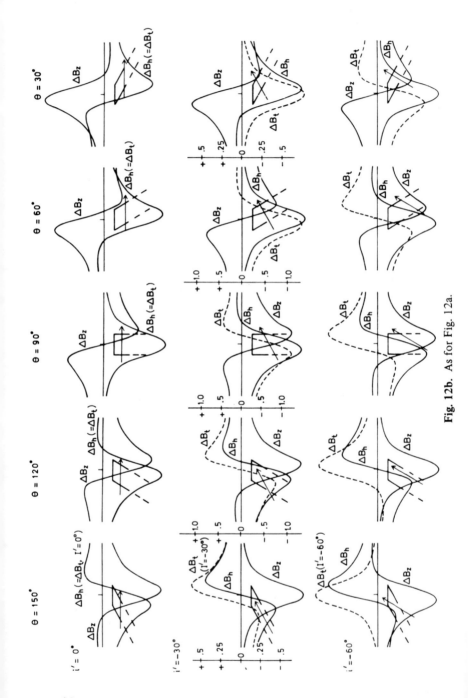

Fig. 12b. As for Fig. 12a.

44

notation is different from that used here the equations given by him are fundamentally exactly the same as (2.38) above. If $b \to 0$, Equation (2.38) reduces to a general expression for the anomaly of a thin sheet, namely

$$\Delta B(l, 0, n) = \left(\frac{\mu_0}{4\pi}\right) 2M'Cb \sin \theta \, \frac{a - kx}{a^2 + x^2} \tag{2.40}$$

C and k are still given by the table above but it should be realized in comparing (2.40) with (2.34), (2.35) and (2.36) that b here is the width of the sheet in the *horizontal* plane and not the width perpendicular to the plane of the sheet as in (2.34) to (2.36). Also, (2.36) is only a special case of (2.40).

The limit leading to (2.40) is easy to obtain but the reader wishing to verify the result will find the following two hints useful:

(1) The difference of the two arctangents in (2.38) must be expressed as a single arctangent by the usual formula

$$\tan^{-1}\alpha - \tan^{-1}\beta = \tan^{-1}\frac{\alpha - \beta}{1 + \alpha\beta} \quad \text{and}$$

(2) r_1^2, r_2^2 must be expressed as

$$(x^2 + a^2) \left\{1 + \frac{b(b/4 \pm x)}{x^2 + a^2}\right\}$$

before proceeding to the limit $b \to 0$.

Many workers have drawn attention to the interesting fact, which the reader can easily verify, that the first x-derivative of ΔB for 'thick' sheets is the sum of the anomalies of two thin sheets coinciding with the lateral surfaces of the thick sheet.

The simplest manner of interpreting magnetic anomalies across geological features resembling thick sheets is to estimate a trial value of k (for instance by assuming M to be in the direction of T_0) and by assuming any reasonable θ. For fairly thick sheets ($b \geqslant \sim 2a$) it is usually a good approximation to put in (2.39) b = distance between the principal inflection points of the anomaly profile and obtain an estimate of a from (2.39). With the values of a, b and k thus estimated ΔB is calculated from (2.38) and the parameters adjusted repeatedly to secure a satisfactory agreement between observed and calculated anomalies. Computer programs can be devised for the purpose. The value of k in the final solution can be used to find θ if this was not known previously. For ΔB_z, for instance, $\theta = i' + \tan^{-1}(k)$ provided demagnetization effects transverse to the sheet are negligible (see further below).

The bracketed part of the right-hand side of (2.37) or (2.38) consists of two terms, one of which $(\alpha_1 - \alpha_2)$ is symmetric in x and the other $(k \ln(r_1/r_2))$ is antisymmetric. This interesting fact has been used [e.g. 16, 17] to construct families of convenient curves for rapid interpretation.

2.10.2.1 *Finite prisms*

Equations (2.37) and (2.38) are valid for sheets extending to infinity in either direction perpendicular to the plane of Fig. 11. The effect of finite depth extent (a two-dimensional prismatic body) can be taken into account by subtracting from the right-hand sides of (2.37) and (2.38) the corresponding express-ions for a sheet of identical width and dip with its upper surface at a depth equal to the depth of the lower face of the prism.

The main effect of a finite depth extent is an enhancement of the side extrema, especially on the down-dip side (cf. Figs. 10c, 13). In Fig. 13 the deeper hole was drilled specifically to verify the inference about the depth extent. The agreement between the interpretation and the drilling results in this example is by no means exceptional but is representative of that achieved in many cases in magnetic prospecting. The magnetic anomalies of prisms whose length as well as depth extent is finite have been studied in great detail by Hjelt [18]. It is evident from this study that the effect of finite strike length is very complex and no simple description of it can be given.

2.10.2.2 *Effect of demagnetization*

It has been shown in Appendix 3 that the magnetizing force within a sphere, on which is acting an external magnetizing force H_0, is $H_0/(1 + \kappa/3)$, $< H_0$ if $\kappa > 0$, because of a demagnetizing force due to the magnetization of the sphere. For bodies of arbitrary shape the internal force may be written as

$$H_i = \frac{H_0}{1 + N\kappa}$$

where N is a demagnetizing 'factor'. It should be noted that only for bodies bounded by second-degree surfaces is N a constant

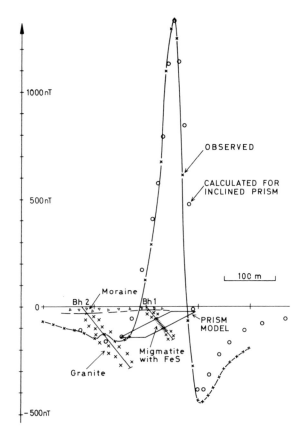

Fig. 13. Observed and calculated ΔB_z across an inclined prism-like body ($k = -1.564$).

within a body (e.g. 1/3 for a sphere). For all other shapes N and hence also H_i and the intensity of magnetization (κH_i) vary from point to point. Consequently an average N is often defined, but it should be realized that there is no unique way of defining N_{av}. Further, N_{av} depends on the direction of H_0. We shall adopt for N_{av} the definition $\int N dv/V$, where V is the volume of the body. Equations for N for rectangular prisms have been published by Joseph *et al.* [19]. A review by Joseph [20] will also be found to be highly illuminating in this context.

Demagnetization effects influence the magnetic interpretation in the following way. The value of k obtained in the final adjustment described above yields $\theta - i'$, the angle between the dip and the inclination of M'. If M' has a remanent as well as

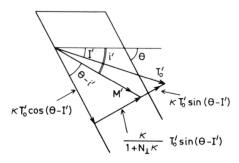

Fig. 14. Effect of demagnetization.

induced component, i' is unknown and hence also θ. However, even when there is no remanence, when M' should be in the direction I', it will deviate from this direction. This is because if the prism is of infinite strike length the magnetization in the cross-sectional plane will not be $\kappa T'_0 \cos(\theta - I')$ but $\{\kappa/(1 + N_{\parallel}\kappa)\} T'_0 \cos(\theta - I')$ where $N_{\parallel}$ is the demagnetization factor in the direction of the depth extent. Similarly the transverse magnetization will be $\{\kappa/(1 + N_{\perp}\kappa)\} T'_0 \sin(\theta - I')$.

For infinite depth extent, as in Fig. 14, $N_{\parallel} = 0$ and $N_{\perp} = 1$ so that $\tan(\theta - i') = \tan(\theta - I')/(1 + \kappa)$. For ΔB_z, for example, $\tan(\theta - i') = k$ and hence

$$\theta = I' + \tan^{-1}\{(1 + \kappa)k\} \tag{2.41}$$

(If unrationalized units are used, κ in (2.41) is to be replaced by $4\pi\kappa$).

For a prism of finite depth extent

$$\theta = I' + \tan^{-1}\left\{\frac{k(1 + N_{\perp}\kappa)}{1 + N_{\parallel}\kappa}\right\} \tag{2.42}$$

It should be observed that the demagnetization effect tends to deflect M' so as to make it less inclined to the sloping sides of the sheet. The neglect of demagnetization can lead to grossly erroneous conclusions about θ if κ is appreciable. Equation (2.42) is also valid for a thin sheet.

2.10.3 Sloping step

It is interesting to note that the anomalies over this feature are again given by (2.37) and (2.38) but with the symbols as in Fig. 15. This figure also shows a sketch of two of the many possible shapes of ΔB_z curves across a step. The shapes of ΔB_h and ΔB_t

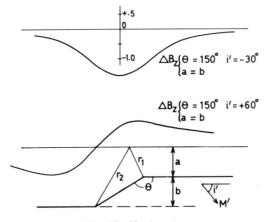

Fig. 15. Sloping step.

curves can likewise take a very wide variety of forms and general rules for determining the parameters of the sloping step are very difficult to formulate. Although θ can be estimated from the value of k that gives the best fit with the observed data, it is not possible to estimate b by any simple rule, contrary to the case for the thick sheet, so that the adjustments in interpretation are much more difficult to make for the sloping step.

2.10.4 Bodies of arbitrary shape

The calculation of magnetic fields ultimately boils down to Equation (2.6) but the integrals involved are generally unmanageable for arbitarily shaped bodies, even for numerical computation on high-speed computers, unless simplifying assumptions are made. One such approach concerns very long features of uniform cross-section. In this case the volume integrals in (2.6) reduce to surface integrals over the cross-section (cf. Appendix 4).

The surface integrals can be converted to line integrals along the boundary of the cross-section by Gauss' well-known theorem and these are easy to evaluate if the boundary is approximated by a polygon with straight sides. The integration then reduces strictly to a summation suitable for computer programming. Shuey and Matthews have generalized this approach to cases where M_x, M_z are linear functions of x and z, and have also given complete computer programs [21]. Two methods analogous to the above have been proposed for three-dimensional bodies by Bott [22]. Both of these can be adapted to programming even on small computers. In one of these the bounding surface S of the body is approximated by polygons. If it is assumed that the magnetization is uniform then the right-hand side of (2.6) becomes the integral of the divergence of $(\mathbf{M}/r)$ and can be converted by Gauss' divergence theorem to a surface integral over S. Bott suggests an ingenious procedure to calculate the contribution of each polygon to the field at the observation point. These contributions are summed to give the desired component of the field. The method of approximating a body by a polyhedron has been extensively treated by Coggon [23] who extends it to gravity calculations as well.

Bott's second method is simpler and faster. It amounts in principle to differentiating (2.6) with respect to ξ, η, ζ under the sign of integration and performing the integration with respect to z. The origin of coordinates is then chosen at Q by putting $\xi = 0$, $\eta = 0$ and $\zeta = 0$. The reader can immediately verify that the magnetizing force in the z-direction, for example, will now be given by

$$Z(0, 0, 0) = \int_{-\infty}^{\infty} \int_{-\infty}^{\infty} - \left[M_x \frac{x}{r^3} + M_y \frac{y}{r^3} + M_z \frac{z}{r^3} \right]_{z=z_1}^{z=z_2} dx \, dy$$

(2.43)

In the computation the body is first subdivided into a number of vertical square (or rectangular) prisms and (2.43) is replaced by a summation over all the prisms:

$$Z(0, 0, 0) = - M_x \Sigma \left(\frac{x}{r^3} \right)_{z_1}^{z_2} \Delta a - M_y \Sigma \left(\frac{y}{r^3} \right)_{z_1}^{z_2} \Delta a - M_z \Sigma \left(\frac{z}{r^3} \right)_{z_1}^{z_2} \Delta a$$

(2.44)

where $\Delta a = dx \, dy$ is the cross-section of a prism whose centre has the coordinates x, y while z_1, z_2 are the distances to the top and base of a prism.

An essentially similar approach in which, however, the integrations in (2.43) have been performed exactly, has been used by Sharma [23] in devising a graphical aid for calculating magnetic anomalies of irregular bodies.

2.11 The Smith rules

The depth to an anomalous structure is a magnitude of prime importance. Some rules for determining it from magnetic observations were given in earlier sections but they presuppose a certain form for the body. A general rule for depth determination does not exist. However, Smith [24, 25] has shown that the maximum depth a at which the upper surface of a body causing a magnetic anomaly may lie, can be determined without making any assumption about the shape of the body.

Let $|\Delta B'_z|_{max}$ and $|\Delta B''_z|_{max}$ denote the absolute maximum values of the first and second horizontal gradients (tesla/m and tesla/m^2) of the vertical flux density anomaly along a measured profile. The magnetization of the body may vary in magnitude but will be assumed to be everywhere parallel though not necessarily in the same sense. Its maximum value will be denoted by $|M|_{max}$ (A/m). Then Smith proves, originally in an unrationalized system but here converted, that

$$a \leqslant 3^{1/2}\pi \left(\frac{\mu_0}{4\pi}\right) |M|_{max} / |\Delta B'_z|_{max} \qquad (2.45a)$$

$$a^2 \leqslant 2^{1/2}6\pi \left(\frac{\mu_0}{4\pi}\right) |M|_{max} / |\Delta B''_z|_{max} \qquad (2.45b)$$

If, however, the more restrictive assumption is made that the magnetization is everywhere vertical and in the same sense, the inequalities are considerably improved and

$$a \leqslant 2.59 \left(\frac{\mu_0}{4\pi}\right) |M|_{max} / |\Delta B'_z|_{max} \qquad (2.46a)$$

$$a^2 \leqslant 3.14 \left(\frac{\mu_0}{4\pi}\right) |M|_{max} / |\Delta B''_z|_{max} \qquad (2.47a)$$

For isotropic bodies without remanent magnetization M_{max} in (2.45) and (2.46) may be replaced by $\kappa B_0 / \mu_0 (1 + N\kappa)$ where B_0 is the normal geomagnetic flux density and N is the demagnetizing factor.

For bodies which extend to great distances perpendicular to the measured profile and in which M is everywhere parallel, the Smith rules are that

$$a \leqslant 4 \left(\frac{\mu_0}{4\pi}\right) M_{max} / |\Delta B'_z|_{max} \qquad (2.47a)$$

$$a^2 \leqslant 1.5\pi \left(\frac{\mu_0}{4\pi}\right) M_{max} / |\Delta B''_z|_{max} \qquad (2.48a)$$

2.12 Some examples of magnetic investigations

The examples below are chosen to illustrate some important points in magnetic interpretation. A general method of attack on any problem is hard to find, especially when igneous rocks or chromite and manganese ore bodies, all of which are notorious for the capricious character of their remanent magnetization, are the objects of investigation. Every case of magnetic investigation needs a careful study of the geology and the topography of the area.

2.12.1 Magnetite ore

The map in Fig. 16a shows the results of a recent survey in Central Sweden. The choice of the area was dictated by general geological considerations; the exact location of the magnetic disturbance is the outcome of the geophysical work. The magnetic anomaly shows an approximately E–W strike. From profiles going over points in the immediate vicinity of the anomaly centre the mean $x_{1/2} : y_{1/2}$ (Section 2.10) was found to be about 0.55 which gave $L/a \approx 3$ and $a/x_{1/2} \approx 1.1$ (Fig. 10c). The mean value of $x_{1/2}$ over the central profiles was 58 m, hence $a = 65$ m. The observed and calculated anomalies using this depth and a magnetic width $(b\kappa) = 86$ also estimated from Fig. 10c are shown in Fig. 16b. A slight effect of the transverse magnetization was also taken into consideration in these calculations. A drill-hole placed as shown encountered rich magnetite ore of total horizontal width of 10 m. This would indicate an average apparent susceptibility of about 8.6 for the ore (cf. Table 1). A hole parallel to the one shown was drilled initially along a line through the anomaly centre, but encountered pegmatite at the expected ore depth and, owing to the peculiar disposition of the pegmatite dike, continued to be in it without giving any ore. This illustrates well the uncertainties that lurk in geophysical work even when the anomaly is a 'text book example' and the agreement between observations and calculations is almost as good as might be desired.

2.12.2 Chromite deposits

Magnetic anomalies over two chromite masses (one known before the work) in the Guleman concession area in Turkey (approximately 39°50′ E and 38°30′ N) are shown in Fig. 17 [26]. According to Yüngül, the susceptibility of the ore masses is, on the average, 2–18 times smaller than the surrounding ultrabasic or basic rocks (serpentines, peridotites, norites) so that negative anomalies should be expected over the ores. This is at variance

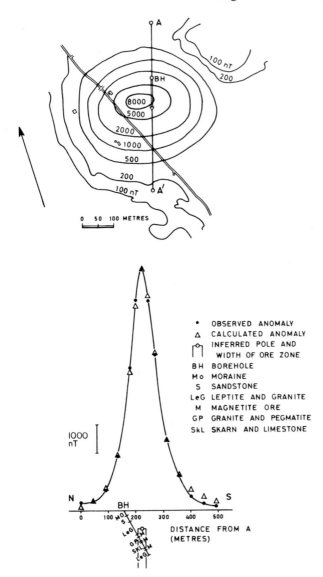

Fig. 16a. A magnetic survey (ΔB_z) in Central Sweden.
Fig. 16b. Profile AA′ in Fig. 16a.

with the observations, the positive values of which must, therefore, be attributed to permanent magnetization pointing downwards. Now both Cr_2O_3 and $FeCr_2O_4$ occurring in chromite ores
are antiferromagnetic, the former with a weak susceptibility. It is,

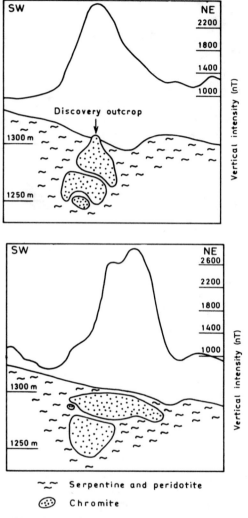

Fig. 17. ΔB_z profiles across two chromite masses [26].

however, conceivable that the latter compound has ferrimagnetic properties so that spontaneous magnetization of the mass may therefore be possible.

2.12.3 Sulphide body near Lam (Bavarian Forest)

The country rock in this area consists mainly of quartzitic shales and evinces a typical layered structure. The dip of the layers is

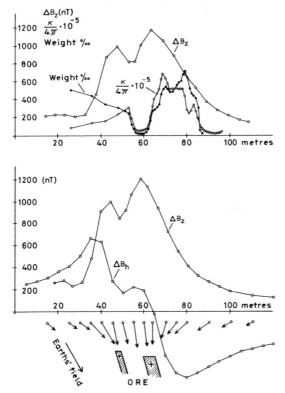

Fig. 18. Magnetic profiles across sulphide veins [27].

about 70–80° towards the north. The ore occurs in an impregnation zone as veins concordant with the shales. It contains pyrite, chalcopyrite, pyrrhotite and galena carrying values in silver. Some magnetite is also present.

Two magnetic profiles over the ore are shown in Fig. 18 after Zachos [27]. Maxima in ΔB_z and inflection points in ΔB_h corresponding to each of the two parallel veins are evident. The arrows in the lower profile represent anomalous total intensity vectors.

In the upper profile are also plotted (1) the susceptibility (κ) (rationalized system) of the rock samples at different places along an underground gallery leading to the ore and (2) the estimated proportion of the total magnetic constituent in the samples. The susceptibility and B_z curves run roughly parallel to each other but the maxima in the former are displaced about 10 m to the north. Zachos attributes this difference to the northerly dip of the veins.

The weight per thousand curve departs from the susceptibility curve at several points. This apparent discrepancy has been attributed to variations in the *magnetite: pyrrhotite* proportion in the samples.

2.13 Measurement of susceptibility and remanence

The intensity of magnetization of rocks can be determined in two basically different ways; either by measuring the field produced by a rock sample or by measuring the effect of the sample on the inductance of an electromagnetic circuit. We shall start with the magnetometer method and for simplicity consider a spherical sample (volume v) placed vertically below a vertical-field magnetometer. If H, Z are the earth's horizontal and vertical magnetizing forces, the induced magnetization intensities are $\kappa_e H$ and $\kappa_e Z$ where $\kappa_e = \kappa/(1 + \kappa/3)$ (Equation (A 3.3), Appendix 3). If the components of the remanent intensity in the sample, M_{rh} and M_{rz}, are directed in the same direction as the induced components, the net horizontal and vertical components of the magnetization intensity will be $\kappa_e H + M_{rh}$ and $\kappa_e Z + M_{rz}$. The components of the magnetic moment are then $(\kappa_e H + M_{rh})v$ and $(\kappa_z + M_{rz})v$.

It follows from Equation (A3.4b), Appendix 3 that directly above the centre of the spherical sample the vertical magnetizing force due to the horizontal moment is zero. The vertical flux density anomaly $\Delta B_z(1)$ read by the magnetometer will be due to the moment $(\kappa_e Z + M_{rz})v$ only and this is easily seen from Equation (2.29) to be

$$\Delta B_z(1) = \frac{\mu_0}{4\pi} \frac{2(\kappa_e Z + M_{rz})v}{a^3} \tag{2.49a}$$

where a = distance between the sample centre and the sensitive element of the magnetometer.

If now the sample is turned upside down, its net magnetic moment in the vertical direction will be $(\kappa_e Z - M_{rz})v$ giving an anomaly

$$\Delta B_z(2) = \frac{\mu_0}{4\pi} \frac{2(\kappa_e Z - M_{rz})v}{a^3} \tag{2.49b}$$

Hence, adding (2.49a) and (2.49b)

$$\kappa_e = \pi \frac{a^3}{v} \frac{\Delta B(1) + \Delta B(2)}{4B_{0z}} \tag{2.50a}$$

where $B_{0z} = \mu_0 Z$ is the normal geomagnetic flux density at the place of measurement.

Similarly, by subtraction,

$$M_{rz} = \pi \frac{a^3}{v} \frac{\Delta B(1) - \Delta B(2)}{4\mu_0} \quad \text{(A m}^{-1}\text{)} \qquad (2.50b)$$

The true susceptibility and remanent magnetization intensity are given by $\kappa_e/(1 - N\kappa_e)$ and $M_{rz}/(1 - NM_{rz})$ respectively, where N is the demagnetization factor ($1/3$ for a sphere).

In low magnetic latitudes ($B_{0z} \approx 0$) a spherical sample directly below the magnetometer will produce practically no anomaly at it in the vertical direction. In this case the sample should be displaced sideways. It can be easily shown from Equations (A 3.4), Appendix 3 that at the magnetic equator itself the optimum displacement is such that the line joining the magnetometer element and the centre of the sphere makes an angle $\tan^{-1}\sqrt{2}$ with the horizontal. Equations (2.49) must, of course, be replaced then by appropriate equations obtained from (2.29), since $x \neq 0$. In any case only samples with relatively high magnetic susceptibility ($\kappa > \sim 1$ if $v \approx 0.5$ litre and $a \approx 1$ m) can be measured in this fashion. For weakly magnetic samples it is necessary to use a more sensitive apparatus, for example an astatic magnetometer. Furthermore, a strong magnetizing force such as that due to a current-carrying solenoid is required instead of the earth's magnetizing force. Werner [28] has described in detail an apparatus for measuring weakly magnetic rock samples.

The principle of the electromagnetic method of measuring the susceptibilities of rocks is to determine the change in the inductance of a solenoid when a magnetic body is introduced into the solenoid. The inductance is usually measured on an alternating current bridge. The change in inductance is related basically to the permeability of the rock material. An apparatus in which the change that results in the mutual inductance of two flat coils when the coils are placed on a smooth outcrop of a rock has been described by Mooney [29]. Apparatuses of this type are also commercially available.

An ingenious, but apparently little known, electromagnetic method with which both the relative magnetic permeability and the electric conductivity of a cylindrical drill core sample can be determined simultaneously has been described by Malmqvist [30]. In this, a long solenoid is fed by an alternating current (frequency $v \sim 10$ kHz) and the flux density ϕ_0 at its centre is measured by a small flat coil. The cylindrical sample (radius r) is then inserted in the small coil, which is made to fit tightly round the sample, and the new flux density ϕ is measured. Actually ϕ differs in phase from ϕ_0 and hence the complex ratio $\phi/\phi_0 = M + iN$ must be measured.

Malmqvist has shown that for a cylinder that is long compared to its radius

$$M = \mu_r [1 - 0.02604(kr)^4 + 0.00162(kr)^8 - \dots] \qquad (2.51a)$$

$$N = \mu_r [- 0.12500(kr)^2 + 0.00629(kr)^6 - \dots] \qquad (2.51b)$$

where $k = (2\pi\mu_r\mu_0 \sigma v)^{1/2}$, σ = electric conductivity (siemen per m) and kr is small (<0.5). Retaining only the first two terms in each expression we can easily obtain kr from the ratio M/N, and inserting this in (2.51a) or (2.52b) find μ_r and hence $\kappa = \mu_r - 1$. Having found μ_r it is a simple matter to obtain σ from the expression for kr. Malmqvist has actually described a graphical method so that the higher terms in (2.51) need not be neglected and has also discussed some of the corrections that must be applied to the measurements.

It should be noted that the electromagnetic methods cannot give the remanent intensity since this, being constant in time, cannot cause any induced electromotive force in the coil. The development of palaeomagnetic research has led to the construction of some highly sophisticated apparatuses for determining the vector of extremely weak remanent intensities in small rock samples. For a discussion of these reference may be made to Collinson *et al.* [31].

3 Gravitational methods

3.1 Introduction

Newton's law of gravitation states that the force (in newton) between two point masses m_1, m_2 is equal to Gm_1m_2/r^2 where r is the distance between the masses and $G = 6.670 \times 10^{-11}$ [$\approx(20/3) \times 10^{-11}$] $m^3/kg\ s^2$. A unit mass placed in the vicinity of any body will be in a field of force (gravitational field) and experience an acceleration. The force may be calculated by applying Newton's law to infinitesimal volume elements of the body and integrating over the entire volume. The earth has also a gravitational field but in calculating it account must be taken of the centrifugal force due to the rotation of the earth.

The earth may be considered to be an ellipsoid of revolution with an ellipticity (equatorial minus polar radius divided by equatorial radius) $= 1/298.2$. The surface of such an ellipsoid of revolution is an equipotential surface. The gradient of the gravitational potential, that is, the force of gravity (g), is by definition everywhere perpendicular to the surface, which means that it acts in the vertical direction. Its variation with the latitude ϕ, at sea level, can be approximated within $1\ \mu m/s^2$ by the formula

$$g = 9.780318(1 + 0.005\ 3024 \sin^2 \phi - 0.000\ 0059 \sin^2 2\phi)\ m/s^2$$

$$(3.1)$$

This formula was adopted by the International Union of Geodesy and Geophysics in 1967 after a critical evaluation of the available absolute g-values in the world [32]. The formula previously adopted by the IUGG in 1930 is still in use for calculating the 'normal gravity' at sea level at any latitude. The difference between the gravitational acceleration values calculated from these two formulas is given to the accuracy quoted above by

$$g_{1967} - g_{1930} = (-172 + 136 \sin^2 \phi) \; \mu m/s^2$$

Some of the most accurate absolute determinations of gravity have been made by means of reversible pendulums of Kater's type. The method consists in principle of adjusting the moments of inertia of a bar pendulum such that its period of vibration (t) about two knife edges located on either side of the centre of gravity are equal. The distance between the knife edges l is then the length of an ideal simple pendulum of the same period so that $g = 4\pi^2 l/t^2$. Other methods of absolute determination have also been devised, e.g. the free fall of a mass or the determination of the paraboloid of revolution obtained by revolving a vessel containing mercury around a vertical axis. The free-fall method has now superseded significantly the accuracy of reversible pendulum measurements [33].

In applied geophysics, a knowledge of the absolute gravity is not of immediate interest. We are concerned, as in the magnetic methods, with relative measurements. These give the gravity difference Δg between an observation point and a base point. Appropriate corrections (Section 3.4) must be applied to the differences measured within any region to reduce them to some standard conditions. The corrected g-values, called the anomalies, yield information about the changes of density within the earth as well as about the surfaces that bound regions of differing density. The information is, however, always subject to certain fundamental ambiguities inherent in the theory of the Newtonian potential (Section 3.8).

Gravity anomalies, being differences in acceleration, can be expressed fundamentally in the SI unit m/s^2 but more conveniently in the sub-unit $\mu m/s^2$. One $\mu m/s^2$ is also called a gravity unit (g.u.). In most current geophysical literature before the adoption of SI the unit gal (cm/s^2) named after Galileo and its submultiple, the milligal are used (1 mgal = 10 g.u.). Since the value of g given by Equation (3.1) varies between the relatively narrow limits of 9.780318 and 9.832177 m/s^2 from the equator to the poles, one g.u. is roughly one ten-millionth (10^{-7}) of the normal gravity at any place on the earth. The maximum gravity anomalies (on the surface) due to concealed features such as salt domes, oil bearing structures, ore bodies, undulations of rock strata etc. are of the order of a few tens to a few hundreds of g.u.

and, in fact, for small-scale or deeply buried structures, they may be only a few g.u. Away from such maxima, the distortions in the normal gravity field of the earth may be even smaller, say, $1-10$ parts in 10^8.

3.2 Gravimeters

It is clear from the above that relative gravity measurements, if they are to have any wide application, must be made with an accuracy better than a few parts in 10^7 and, preferably, with an accuracy approaching $1-5$ parts in 10^8. This aim is achieved in instruments known as *gravimeters*. A number of ingenious gravimeter designs have been proposed during the last fifty years but fundamentally they fall into only two categories, the stable and the unstable types. To this may be added a third type, namely the dynamic, but this has seldom been used for geophysical purposes. The stable gravimeter can be described briefly as a highly sensitive balance. It contains a responsive element, usually a spring carrying a weight, which is displaced from the equilibrium position when the force of gravity changes. The displacements are always extremely small (of the order of a few tenths of a nanometre) and must be magnified optically, mechanically or electrically. The unstable gravimeter is designed so that when its sensitive element is displaced due to a change in the gravity, other forces tending to increase the displacement come into play. The gravity change can be measured by the force necessary to return the element to its equilibrium position. Brief descriptions of some gravimeter designs are given below to illustrate the principles. More detailed descriptions of these and other gravimeters may be found elsewhere[34].

3.2.1 Stable types

3.2.1.1 *Askania*

In this instrument (Fig. 19a) a beam carrying a mass at one end is held horizontally by means of a main spring (S). A mirror placed on the mass reflects a light beam into a double photoelectric cell. The movement of the mass due to a change in the gravity is indicated by the deflection of a galvanometer through which the differential current from the photoelectric cell is led. The mass is restored to the equilibrium position by varying the tension in the auxiliary spring (S'). Calibration can be effected by means of small known weights brought on the beam by tilting the instrument.

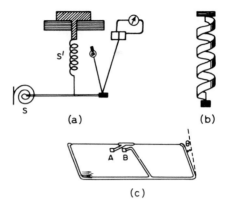

Fig. 19. Stable gravimeters.

3.2.1.2 *Gulf (Hoyt)*

This gravimeter utilizes a helical spring formed from a ribbon (Fig. 19b). One end of the spring is rigidly clamped while the free end carries a mass with a mirror. An elongation of a helical ribbon spring is always accompanied by a rotation of the free end. In the Gulf gravimeter the rotation is much greater (and therefore can be read more accurately) than the elongation (or the contraction) of the spring caused by a change in the gravity. The range of the instrument is only about 300 g.u. so that a readjustment of the tension in the helix is necessary if gravity differences larger than this amount are to be measured. The accuracy is of the order of 0.2–0.5 g.u.

3.2.1.3 *Nörgaard*

This is one of the gravimeters (Fig. 19c) combining a wide range (about 20 000 g.u.) with a relatively high accuracy (about 1 g.u.) A small quartz beam carrying a mirror A is supported horizontally from a quartz thread, the torsion in the latter counteracting the force of gravity. The mirror A is initially parallel to the fixed mirror B as is indicated by the coincidence of two index lines in the field of a telescope. When the beam deflects due to a change in the gravity, coincidence can be achieved again by tilting the entire frame through an angle θ. There are two such positions of the frame, one on each side of the initial position. At coincidence the torsion moment of the thread must always be the same ($mg_0 l$) so that if g and g_0 are the gravity values at two stations then $g \cos \theta = g_0$.

The instrument can be calibrated by tilting it at small known angles.

3.2.2 Unstable types

3.2.2.1 *LaCoste–Romberg*

This gravimeter is essentially an adaption of the long period LaCoste seismograph [35, 36] which uses a 'zero-length' spring. Such a spring is wound so that its extension is equal to the distance between the points at which its ends are fastened. Thus, the length defined as the actual physical length minus the extension is zero. The zero length spring S (Fig. 20a) is attached rigidly to the frame at C and balances the mass M at the end of a beam. With the geometry as in the figure, it is easy to show that the net torque on the mass is $(Mg \times \overline{AM} - k \times \overline{AB}^2)\sin \theta$ where k is the spring constant. If $Mg \times \overline{AM} = k \times \overline{AB}^2$ the torque becomes zero, the period infinite and the equilibrium unstable. The

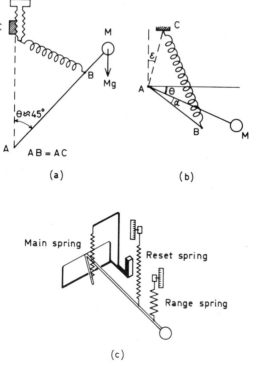

(a)   (b)

(c)

Fig. 20. Unstable gravimeters.

instrument is then very sensitive to variations in *g*. Readings are taken by restoring *M* to the original position by raising or lowering *C* by means of a screw with a calibrated dial. The accuracy is of the order of 0.2 g.u.

3.2.2.2 *Worden*

The principle of this instrument is very similar to that of the LaCoste–Romberg gravimeter. The mass *M* (Fig. 20b), is kept in unstable equilibrium by the zero length quartz spring, *BC*, whose one end is attached to an arm *AB* inclined at a fixed angle α to *AM*. Both *AM* and *AB* are hinged at *A* to a torsion thread. If θ is the deflection of *AM* from the horizontal the net torque on the beam system is easily seen (on applying the elementary sine theorem to the triangle *ABC*) to be $(Mg\overline{AM} \cos\theta - \tau(\theta + \theta_0) - k\overline{AC} \times \overline{AB} \cos(\alpha + \theta - \epsilon)$ where θ_0 is the permanent torsion in the quartz thread and τ is the torsion constant.

By suitable choice of the different constants and the position of $C(\epsilon \approx 0)$, the equilibrium can be made unstable and the system becomes very sensitive to variations in *g*. The equilibrium is restored by means of auxiliary springs arranged as shown in Fig. 20c, one of which determines the range of *g* measurable by the instrument and the other compensates for the variations in *g* for a particular setting of the range spring. The instrument is temperature compensated by auxiliary quartz springs and moreover the entire system, except for the reading dials, levels, etc. is kept in a small sealed thermos flask. The total weight of the instrument including the case is about 5 kg but the mass *M* (made out of fused quartz) weighs only a few milligrams. The accuracy is 0.1–0.2 g.u. and the range of the instrument is wide, namely about 20 000 g.u.

3.3   Field procedure

Gravimeter observations are usually made at the corners of a square grid. The length of the side of the square will depend upon the anticipated dimensions of the features to be located. In oil prospecting the grid side may be of the order of 0.5–1 km or more, while in mineral exploration the stations must often be spaced on a grid with sides no larger than 10–50 m. For other purposes, such as the location of dikes or geological faults, the spacing may be anywhere between these extremes. In large-scale or regional surveys it is also the general practice to establish gravimeter stations along roads.

The geographical positions and the elevations of gravimeter stations must be accurately known in order to reduce the readings to standard reference conditions, as described below. The elevations may be determined by spirit-levelling or barometrically [37].

The readings of all gravimeters drift more or less with time, due to elastic creep in the springs. This apparent change in the gravity at a station may be from a few tenths of a g.u. to about 10 g.u. per hour. In order to correct for it, the measurements at a set of stations are repeated after 1–2 h and the differences obtained are plotted against the time between two readings at a station. A 'drift curve' can then be drawn and the corrections read off it. In accurate work it is advisable to determine the drift curve by a least squares adjustment. This is usually straightforward since most gravimeters drift linearly with time. Parabolic or other drift functions are, however, not uncommon.

3.4 Corrections to gravity observations

The gravity difference between two stations is in part due to factors other than the attraction of unknown anomalous masses. These factors and the corrections due to them are as follows.

3.4.1 Latitude

The value of gravity increases with the geographical latitude. By differentiating Equation (3.1) we get

$$\frac{\mathrm{d}g}{\mathrm{d}\phi} = 51\,859 \sin 2\phi \text{ g.u./rad}$$

If the latitude difference between two stations is small the correction becomes

$$\delta g = 0.081 \sin 2\phi \text{ g.u. per 10 m (north–south)} \qquad (3.2)$$

since the mean radius of the earth is $R = 6368$ km.

It must be subtracted from or added to the measured gravity difference according to whether the station is on a higher or lower latitude than the base station. The correction is linear for distances in the N–S direction of the order 1–2 km (about 0.5–1 min of latitude) on either side of the base. If the measurements extend beyond this distance a new base station must be selected and the difference between the normal gravity at it and the first base must be determined by reference to Equation (3.1).

If the north–south distance of a station from the base is known to within 10 m, an accuracy which it is normally not in the least difficult to achieve, the latitude correction will be known to better than one tenth of a g.u.

3.4.2 Elevation

The force of gravity outside the earth varies in inverse proportion to the square of the distance from the earth's centre. If g_1 be its value at the datum level (not necessarily the sea level), then at a height h above it,

$$g = g_1 \frac{R^2}{(R + h)^2} \approx g_1 (1 - 2h/R) \tag{3.3}$$

if powers of h/R higher than the first are neglected. In most practical cases the distance of the datum level from the earth's centre may be taken to be equal to the mean radius of the earth. The correction for the elevation (the 'free-air correction') is then

$$\delta g = \frac{2g_1}{R} h = 3.072h \text{ g.u. } (h \text{ in metres}) \text{ at the equator}$$
$$= 3.088h \text{ g.u.} \qquad \text{at the poles} \tag{3.4}$$

The mean amount 3.080 g.u. per metre of elevation, which is sufficiently accurate for most purposes, must be added to a measured gravity difference if the station lies above the datum level and subtracted if it lies below it.

If an accuracy of 0.1 g.u. is aimed at in relative gravity measurements, elevation differences from the datum level must be known to better than 4 cm.

3.4.3 Material between station levels

It will be realized by reference to Fig. 21a that while the gravity at B will be less than that at A by an amount $2g_1 h/R$ (free-air correction), it will be greater by an amount $\delta g = 2\pi G \rho h$. This is the additional attraction exerted on a unit mass by the slab of rock material of density $\rho(\text{kg/m}^3)$ between the levels of A and B. The correction

$$\delta g = 0.4191 \times 10^{-3} \rho \quad \text{g.u. per metre (of elevation)} \tag{3.5}$$

is called the Bouguer correction. It must be subtracted from the measured gravity difference if a station lies above the level of the base station and added if it lies below.

When measurements are made below the earth's surface the slab of material between A and B (Fig. 21b) exerts an attraction on a

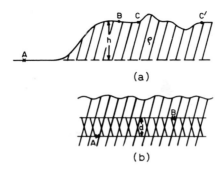

Fig. 21. Rockmass between station and plane of reduction.

unit mass placed at A as well as B. These attractions being in opposite directions, the difference of gravity between A and B due to the slab is $4\pi G\rho d$ and the Bouguer correction is doubled ($0.8382 \times 10^{-3} \rho$ g.u. per metre).

3.4.4 Topography

At a point such as C, a topographic irregularity (hill, knoll, slope, etc.) will exert an attraction directly proportional to its density. The vertical component $(T\rho)$ of this attraction will be directed upwards and reduce the gravity at C. A term of this magnitude must therefore be added to the measured value of gravity at C. A valley such as that near C' is a negative mass and the vertical component of its attraction will also be directed upwards leading again to an additive topographic correction.

The topographic correction is often calculated by dividing the area around a station in compartments bounded by concentric rings and their radii drawn at suitable angular intervals (ϕ). The mean elevation (z) in each compartment is determined from a topographic map, without regard to sign, that is by treating a hill as well as a valley as positive height difference from the station level. The correction due to the attraction of the material in such a compartment is

$$\delta g = T\rho = G\rho\phi[r_2 - r_1 + \sqrt{(r_1^2 + z^2)} - \sqrt{(r_2^2 + z^2)}] \qquad (3.6)$$

where r_1, r_2 are the radii of the inner and outer rings bounding the compartment.

Tables of the bracketed expression in (3.6) have been published, but with the general availability of small programmable calculators it has become easy, and is in fact preferable, to prepare one's own tables by choosing r_1, r_2 values that best suit the survey and the terrain in question. The calculation of terrain corrections is very tedious and it is therefore fortunate that they are not needed as a rule except for stations in the immediate vicinity of a 'violent' topographic irregularity. There is no compelling reason for a circular division of the terrain except some convenience in *manual* calculations. For computer-adapted procedures a rectangular grid division is preferable. In a method suggested by Ketelaar [38] the terrain surface is approximated by square prisms with sloping upper surfaces. The correction due to a prism characterized by the matrix indices i, j, and surface slope α, is shown to be

$$\delta g(i, j) = G\rho D(1 - \cos \alpha)K(i, j)$$

where D is the grid side. The matrix $K(i, j)$ can be calculated once for all.

3.4.5 Tides

The attractions of the sun and the moon may change the gravity at a station cyclically with an amplitude of as much as 3 g.u. during the course of a day. The correction cannot be calculated in any simple way and recourse must therefore be had to tables regularly published in advance for each year [39]. The drift correction to gravimeter readings includes, in part, the tidal correction.

3.5 Marine gravity measurements

Gravimeter measurements at sea are necessarily more complicated because no stable platform is available. Nevertheless instruments for the purpose have been devised, the earliest one being a three-pendulum apparatus of Vening Meinesz from the 1920's. This is no longer used. Present measurements are made by gravimeters mounted on gyro-controlled stablized platforms. There are a variety of such instruments, some of which have been described by Dehlinger and Chiburis [40].

Gravity measurements on board ship are affected by the earth's attraction as well as by ship motion. The effects of horizontal and vertical accelerations due to the ship's motion and the cross-coupling effects between these two vary from one instrument design to another and are computer corrected in modern instruments. But apart from these, and apart from the corrections described for land work, there is a new, fundamental correction in gravity work at sea explained below.

The centrifugal acceleration experienced by a mass, at rest relative to the earth's surface at latitude ϕ, due to the earth's rotation from west to east is $V_\phi^2/(R \cos \phi)$ where V_ϕ is the linear velocity of the mass due to the earth's rotation. Its component in the radial, that is, vertical direction (V_ϕ^2/R) tends to reduce the earth's gravitational pull. In land work this is included in the net measured value of g. For an instrument on board a ship moving with easterly velocity V_E the centrifugal acceleration is $(V_\phi + V_E)^2/(R \cos \phi)$ and its vertical component is $(V_\phi + V_E)^2/R$. The vertical acceleration due to a northerly speed V_N is, however, V_N^2/R. The additional acceleration due to the ship's motion is then easily seen to be $(2V_\phi V_E + V^2)/R$ where $V = (V_E^2 + V_N^2)^{1/2}$ is the ship's speed. Since $V_\phi = R\omega \cos \phi$ where ω is the earth's angular velocity the acceleration can be written as

$$\delta g_E = (75 V_E \cos \phi + 0.04 \ V^2) \quad \mu\text{m/s}^2$$

where V and V_E are in knots.

The correction δg_E is called the *Eötvös correction* and is to be added to the measured value of gravity. (V_E is negative for a westerly speed.) Some gravity work under water is done by lowering a gravimeter enclosed in a pressure-tight spherical shell onto the sea bottom. Levelling of the instrument is executed by remote control devices. The Eötvös correction does not, of course, arise in such measurements. The overall accuracy of sea gravity measurements is between about 10 and 100 $\mu\text{m/s}^2$ depending upon errors in ship speed, ship heading, positioning etc. For the LaCoste–Romberg ocean bottom gravimeters, however, an accuracy of 1 $\mu\text{m/s}^2$ and better has been claimed.

3.6 The Bouguer anomaly

It will be seen now that the corrected gravity difference between a station and a base in land measurements is

$$\Delta g_{corr} = \Delta g_{obs} + 3.080 \ h - 0.4191 \times 10^{-3} \ h\rho + T\rho \quad \text{g.u.} \tag{3.7}$$

where h is positive if the station is above the base and negative if it is below. The latitude and tidal corrections are included in the term Δg_{obs}. The density of the topographic irregularity in Fig. 21 is assumed to be the same as that of the infinite slab, namely ρ. This assumption may not be always justified. The numerical coefficient of the third term on the right-hand side must be doubled if measurements are made underground. In such measurements T may be negative if, for instance, tunnels are situated *above* the level of a station.

If Δg is to be expressed in mgal the coefficients of h and $h\rho$ in

(3.7) must be divided by 10. If, in addition, ρ is in g/cm^3 the third and fourth terms must be multiplied by 1000.

The difference Δg_{corr} is called the (relative) Bouguer anomaly. Its variations are to be attributed to the variations of density below the datum level implied in Equation (3.7) provided the Bouguer correction also includes the attraction of any inhomogeneities in the slab of material between the surface and the datum level. Since this situation is not realized in practice, the Bouguer anomaly represents the effect, at points on the physical surface of the earth, of all inhomogeneities within the earth. The reader should also guard himself against the loose expression that the Bouguer anomaly is the 'gravity reduced to the datum level'. It should be realized that the gravity measured on the datum level, at a point B', vertically below B in Fig. 21a, will *not* differ from that at B by Δg_{corr}, because the effect of any inhomogeneities below the datum level is not the same at B as at B'.

3.7 Density determinations

Strictly speaking, any value may be chosen for ρ for the reductions in Equation (3.7). However, it is clearly desirable to eliminate the effect of surface features as far as possible by using the true (average) value for their density.

The density may be estimated by laboratory measurements on samples of the rocks exposed within the area of interest. But such estimates suffer from the fact that the samples may be weathered or in other ways unrepresentative. Furthermore, the density of the rocks at depth may be different from that of the surface samples, owing to a variable water content and, in the case of 'loose' rocks like clays, marls, moraine, etc., owing to significant compaction even at moderate depths. Therefore various 'field' methods have been suggested for the determination of ρ. In Nettleton's method [41], the Bouguer anomalies at the stations on a line of measurement are calculated assuming different values of ρ. The anomalies that show the least correlation with the topography are adopted as the true anomalies, the corresponding ρ being adopted as the true average density of the surface rocks. Nettleton's original method is graphical, but Jung [42] has pointed out that it can be translated into exact mathematical language by putting the correlation coefficient between Δg_{corr} and h equal to zero. We then get

$$\rho = \rho_0 + \frac{\Sigma \left(\Delta g_{corr} - \overline{\Delta g_{corr}}\right)(h - \overline{h})}{0.0004191 \, \Sigma (h - \overline{h})(h - \overline{h} + T - \overline{T})} \tag{3.8}$$

where ρ_0 is an appropriate assumed value of ρ to which a 'correction' term must be added to obtain the true ρ.

In a method proposed by Parasnis [43], $\Delta g_{obs} + 3.080\,h$ in Equation (3.7) is plotted against $(-0.0004191\,h + T)$ and the slope of the straight line (determined by least squares) is adopted as the true ρ. This is equivalent to assuming Δg_{corr} (the Bouguer anomaly) to be a random error. In this case

$$\rho = \rho_0 + \frac{\Sigma(\Delta g_{corr} - \overline{\Delta g_{corr}})\,(h - \bar{h} + T - \bar{T})}{0.0004191\,\Sigma(h - \bar{h} + T - \bar{T})^2} \tag{3.9}$$

This is essentially a generalization of a method due to Siegert [44] in which the topographic correction is neglected. Jung has pointed out that while (3.8) and (3.9) reduce to identical forms if $T = 0$, the two values differ but little (a few parts in 1000) in an actual case even if $T \neq 0$. Equation (3.9) and the associated equation for determining the standard deviation of ρ are considerably simpler from the computational point of view than (3.8).

Legge [45] has described a method (neglecting T) in which Δg_{corr}, instead of being treated as a random error, is developed as a power series in the distance of a station from the base.

The above methods deal with stations along a 'line'. The density obtained from (3.8) or (3.9) is then strictly speaking valid only for these stations and the intermediate points. Legge, in the above paper, and Jung [46] have shown that the least squares method leading to (3.9) can be generalized to apply to a set of measurements over an area also.

3.8 Interpretation

The interpretation of gravity anomalies in terms of subsurface mass distributions generally follows a pattern similar to the interpretation of magnetic anomalies (Chapter 2). The mass distribution is assumed to correspond to some plausible simple structure and the parameters of the structure are adjusted until its calculated anomaly at all points agrees satisfactorily with the observed anomaly. For this purpose we need to know the gravity anomalies produced by a variety of type-structures (Fig. 22). Some formulae for such anomalies will be given below without proof. Their derivation is generally elementary. We shall hereafter denote the Bouguer anomaly (Δg_{corr}) by Δg. The anomalies in gravity arise from relatively small density differences (δ) between rock formations and the interpretation is naturally very sensitive to the available density values. It is, therefore, important to obtain

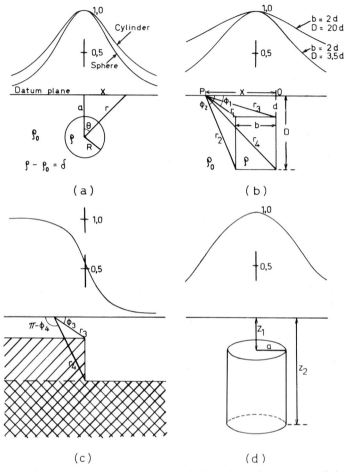

Fig. 22. Gravity anomalies across some bodies of simple geometrical shape.

reliable density estimates of the rocks in an area, preferably by a combination of the laboratory and field methods of the previous section. The densities of some commonly encountered rocks and minerals will be found in Table 6.

3.8.1 Regionals and residuals

Very frequently in gravity work anomalies varying slowly with distance are found to be superimposed on faster variations. The former are termed regional and the latter local variations. Either may be the object of detailed interpretation and is then separated from the other, in which case it is called *residual* variation. More

Table 6   Densities (kg/m^3)

Oil	900	Granite	2500–2700
Water	1000	Anhydrite	2960
Sand, wet	1950–2050	Diabase	2500–3200
Sand, dry	1400–1650	Basalt	2700–3200
Coal	1200–1500	Gabbro	2700–3500
English Chalk	1940	Zinc blende	4000
Sandstone	1800–2700	Chalcopyrite	4200
Rock salt	2100–2400	Chromite	4500–4800
Keuper marl	2230–2600	Pyrrhotite	4600
Limestone		Pyrite	5000
(compact)	2600–2700	Haematite	5100
Quartzite	2600–2700	Magnetite	5200
Gneiss	2700	Galena	7500

commonly it is the local variations that are of immediate interest, and the term residual is commonly understood to imply these.

It is important to realize from the outset that there is no unique way of separating regionals and residuals on the basis of gravity data alone, analytically or otherwise. The reason is simply the obvious fact that a given number cannot be uniquely decomposed into two or more numbers whose sum it is unless extraneous constraints are placed. Nevertheless, for the sake of speed and uniformity in treating the data, it is often desirable to use some standardized procedure to separate different types of gravity variations.

One method is to take the average of a set of symmetrically placed values around a point, subtract it from the value at the point, and prepare a map of the residuals thus obtained. A more sophisticated procedure is to express the gravity field in the area by a low-order polynomial, e.g.,

$$\Delta g(x, y) = Ax + By + 2Cxy + Dx^2 + Ey^2 + F$$

and determine the coefficients $A, B \ldots$ etc. by a least squares adjustment. The residuals are obtained by subtracting the polynomial values from the measured ones. The polynomial, being of low order, represents slow or large-scale variations which are usually attributed to deeper causes, although this need not be their origin, a point that is easily overlooked.

Still more sophisticated procedures using Fourier transformation of the data have also been suggested [47]. In the other extreme, graphical methods in which contours or profiles are smoothed freehand to separate a smooth trend and local variations may be found to be adequate in many cases.

3.8.2 Anomalies of different bodies

3.8.2.1 *Sphere*

With the notation of Fig. 22,

$$\Delta g = \frac{4}{3} \pi R^3 G \delta \frac{\cos \theta}{r^2}$$

$$= \Delta g_{max}/(1 + x_a^2)^{3/2} \tag{3.10a}$$

where $\Delta g_{max} = \Delta g$ at $x = 0$ and $x_a = x/a$. It should be noted that even if the cause of an anomaly is *known* to be a spherical body, its radius and density difference from the surrounding rock cannot be separately determined with the knowledge of the gravity field alone. This is because all spheres with the same value for the product $R^3 \delta (R < a)$ and having the same centre produce identical gravity fields at the surface.

As an example of an approximately spherical body we may take a boulder. A boulder of 1 m radius is buried in an otherwise homogeneous moraine with which it has a density contrast $\delta = 1000$ kg/m^3. If it just touches the surface of the ground, the gravity anomaly directly above its centre will be 0.28 g.u.

3.8.2.2 *Horizontal cylinder*

We take Fig. 22 to represent an infinitely long cylinder striking at right angles to the plane of the paper. Along a line on the ground, perpendicular to the cylinder,

$$\Delta g = \frac{2\pi G R^2 \delta a}{r^2}$$

$$= \frac{\Delta g_{max}}{(1 + x_a^2)} \tag{3.10b}$$

where

$$\Delta g_{max} = 2\pi G R^2 \delta/a.$$

The attraction of an infinitely long cylinder at points outside it is the same as that of an infinitely long rod with the same mass per unit length $(\pi R^2 \delta)$. Note that Δg_{max} decreases as the first power of the depth to the cylinder axis.

As an example we may note the following. A long, horizontal underground tunnel of circular cross-section (radius 1 m), driven in a rock with density 2700 kg/m^3 will produce a maximum decrease of 0.113 g.u. in the gravity at the surface if the axis of the tunnel is at a depth of 10 m.

3.8.2.3 *Rectangular prism*

The cross-section of an infinitely long prism striking perpendicular to the plane of the paper is shown in Fig. 22b. The angles made by the lines $r_1, r_2, \ldots$, etc. with the horizontal are denoted by $\phi_1, \phi_2, \ldots$, etc. We have then

$$\Delta g = 2G\delta \left[x \ln \frac{r_1 r_4}{r_2 r_3} + b \ln \frac{r_2}{r_1} + D(\phi_2 - \phi_4) - d(\phi_1 - \phi_3) \right]$$

(3.10c)

This formula can be derived in a very simple manner (see p. 96).

The rectangular prism is one of the versatile models which can be used to approximate many geological features. If we imagine the lower boundary face to be extended indefinitely in the horizontal plane, the figure would represent a buried ridge or an undulation of the bedrock. If the vertical faces be moved in opposite directions to infinity, $r_1 \rightarrow r_2, r_3 \rightarrow r_4, \phi_1$ and $\phi_2 \rightarrow \pi, \phi_3$ and $\phi_4 \rightarrow 0$ so that Equation (3.10c) reduces to

$$\Delta g = 2\pi G\delta(D - d) ,$$

(3.10d)

which is the anomaly of an *infinite slab* of thickness $(D - d)$. It depends only on the thickness and not on the depth to the top surface of the slab.

If the lower surface is moved downwards to a very large distance, the figure would correspond to a vein or a dike. In this case $r_2 \rightarrow r_4 \approx D$ and $D(\phi_2 - \phi_4) \rightarrow b$ in (3.10c). The anomaly of a dike increases indefinitely with its depth extent since $r_2 \rightarrow \infty$ in the second term but the increase is very slow.

The step structure shown in Fig. 22c is a special case in which one of the vertical faces of the prism is moved to infinity. The prism is then represented by the singly-hatched portion of the diagram. The anomaly of this configuration ($r_1 \rightarrow r_2, \phi_2$ and $\phi_1 \rightarrow \pi$) is the same whether the density of the underlying material (the doubly-hatched portion) is ρ or ρ_0 or has any other uniform magnitude. In the first case, (ρ), we have either a buried erosional *escarpment* or alternatively a vertical *geologic fault* in which the rock strata on one side of a vertical plane have been thrust downwards or upwards with respect to those on the other. The 'throw' of the fault is $(D - d)$. The other two cases resemble *flat-lying features* (laccoliths, ore-beds, etc.) with very large dimensions perpendicular to the plane of the figure.

3.8.2.4 *Vertical cylinder*
The anomaly of a vertical cylinder (Fig. 22d) at a point P_0 on its axis is given by

$$\Delta g = 2\pi G \delta [z_2 - z_1 + (z_1^2 + a^2)^{1/2} - (z_2^2 + a^2)^{1/2}] \qquad (3.10e)$$

If $z_2 \to \infty$,

$$\Delta g = 2\pi G \delta [(z_1^2 + a^2)^{1/2} - z_1] \qquad (3.10f)$$

and if, in addition, $z_1 \to 0$

$$\Delta g = 2\pi G \delta a \qquad (3.10g)$$

These equations show that the attraction of a vertical cylinder remains finite as its depth extent increases, in contrast to the attraction of an infinitely deep-going dike.

The attraction of the cylinder at a point P off the axis is often estimated as follows. The attraction of a cylinder of thickness $(z_1 - z_2)$ and radius $(x - a)$ having the vertical line through P as the axis, is subtracted from the attraction of coaxial cylinder of radius $(x + a)$. The result (valid for $x \geqslant a$) is multiplied by

$$\frac{\pi a^2}{4\pi x a} = \frac{a}{4x}$$

which is the ratio of the area of the plane face of the cylinder in Fig. 22d to the area of the annulus between the two cylinders with the axis through P. For $x < a$, the attraction is estimated by a somewhat more complicated by essentially similar procedure. The method is approximate, however, since the underlying assumption that the anomaly of a cylinder is proportional to the area of its plane face is valid only if the horizontal distance of P is large. The exact expressions for the attraction at P are complicated and can be derived by using elliptic integrals [48].

3.8.2.5 *Bodies of arbitrary shape*
A surprisingly large number of geological structures can be adequately represented by the above regular shapes and their combinations. Formulae for the gravity anomalies of other regular shapes involving dips or sloping faces have also been derived. Often, however, it is required to calculate the attraction of

irregularly shaped bodies. For this purpose a number of graphical methods can be used [49, 50]. One of them, due to Lindblad and Malmqvist, is illustrated by Fig. 23. The graticule in the figure is so constructed that each field, representing a element of the cross-section of a 100-m long prism, contributes 0.1 g.u. to the gravity field at *P*. The density of the prism is assumed to differ by 1000 kg/m^3 from that of the surrounding rock. The prism strikes perpendicular to the plane of the figure 50 m on each side. If a closed curve is traced on the diagram, the anomaly at *P* due to the prism with this curve as the uniform cross-section is found by counting the number of fields within the curve. By shifting the graticule and the curve with respect to each other, the anomaly at any point situated in the vertical plane through the central cross-section of the prism may be obtained.

A more general graphical method is the dot chart of Morgan and Faessler [51] constructed on similar principles but capable of being used for bodies of any length in the strike direction. Nowadays, the calculations are often made using digital computers by approximating the irregular cross-section of the body by a closed polygon, and the body by a number of such polygons on

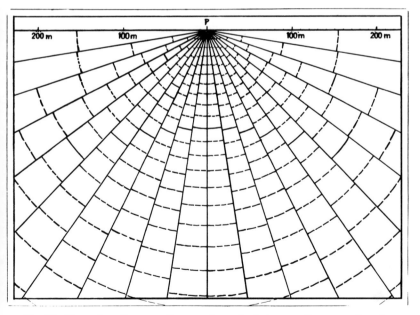

Fig. 23. Example of sector diagram for calculating gravity anomaly of a horizontal prism of uniform cross-section [49].

one another [52]. The vertices of the 'best' polygon can be determined by means of optimization techniques (Chapter 4) starting from an arbitrary polygon. If the sides of the polygon are chosen to be parallel to the x- and z-axes, a much quicker, though slightly less versatile method, described on page 96, results. The approximation, however, will be found to be adequate for most practical purposes.

3.9 Limitations on gravity interpretation

3.9.1 Ambiguity in interpretation

The discussion in the last section may be called the 'forward' approach to gravity interpretation. That is, given a mass distribution we determine its gravity field on the earth's surface. The real problem in applied geophysics is, however, the inverse one: given a gravity field g over an infinite horizontal plane, to determine the mass distribution producing the field. The problem of magnetic interpretation (p. 28) is essentially the same because here too we are concerned with a field of force derivable from a Newtonian potential. Therefore the following discussion applies to magnetic anomalies also.

Suppose we are given a Newtonian potential function $U(x, y, z)$.

$$g = -\frac{\partial U}{\partial z} \tag{3.11}$$

Let the masses producing the potential lie entirely below some plane A (Fig. 24), not necessarily the earth's surface. Then in the space R above A, the potential U is harmonic ($\nabla^2 U = 0$). Let P be any given point and Q a second point in R. Then $1/r = 1/PQ$ is also a harmonic function throughout R if we exclude the region within a small sphere Σ surrounding P. Applying Green's theorem in its symmetric form to U and $1/r$ in the space R, we get

$$\iint_{R-\Sigma} \left(U \frac{\partial}{\partial z} \frac{1}{r} - \frac{1}{r} \frac{\partial U}{\partial z} \right) dS + \iint_{\Sigma} \left(U \frac{\partial}{\partial n} \frac{1}{r} - \frac{1}{r} \frac{\partial U}{\partial n} \right) dS = 0 \tag{3.12}$$

It is not difficult to show that as $\Sigma \to 0$, the second integral in (3.12) reduces to $4\pi U(P)$ where $U(P)$ is the potential at P due to the masses below A (see Appendix 1). Hence,

$$U(P) = \frac{1}{4\pi} \iint_{A} \left(\frac{1}{r} \frac{\partial U}{\partial z} - U \frac{\partial}{\partial z} \frac{1}{r} \right) dS \tag{3.13}$$

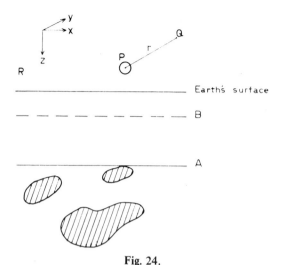

Fig. 24.

The surface integral reduces simply to that over the boundary A of R since the integrand vanishes at infinity. But this equation shows that $U(P)$ is the combined potential of a mass distribution of surface density $-(1/4\pi G)(\partial U/\partial z)$ and a double distribution (dipoles) of moment $U/4\pi G$, over the plane A. Obviously, if the masses below A were replaced by these two distributions on A, the potential at any point above A will be unaffected. Since U is also harmonic in the subspace above any other arbitrarily chosen plane B, shallower than A, the same reasoning will hold and $U(P)$ will be unaltered if the masses are now replaced by a suitable distribution on B. The inverse problem of gravity interpretation has thus no unique solution.

It might be objected that, in reality, we are measuring only the gravity $g = -(\partial U/\partial z)$ so that U in the second term in (3.13) is unknown. The objection is only apparent for, by setting up a Green's function of the second kind*

$$H = \frac{1}{r} + \frac{1}{r'}$$

where r' is the distance of Q from the mirror image of P in the earth's surface, it can be proved that

$$U(x,y,z) = \frac{1}{2\pi} \iint \frac{1}{r} g(x_0, y_0, 0) \, \mathrm{d}S \tag{3.14}$$

---

*That is, a function whose derivative normal to a boundary vanishes on the boundary.

In Equation (3.14) x, y and z are the coordinates of P and $x_0, y_0, 0$ of a point on the earth's surface at which the value of gravity is g. Thus, when g is given at all points on the earth's surface the potential can be found not only at a point on the surface ($r = Q_0 C$, Fig. 25) but at any point in the space above. In practice, the integration will be replaced by a summation where it is assumed that g is the mean value of the gravity within a suitably chosen areal element dS.

Equation (3.14) shows that if the masses producing $g(x_0, y_0, 0)$ are replaced by a thin layer of surface density $g(x_0, y_0, 0)/(2\pi G)$ the potential in the space above the masses will not be altered in any way. Such a layer is known as *Green's equivalent stratum*.

Another important consequence of (3.14) is that the derivative of U at x, y, z in *any* direction l can be uniquely determined by the measured distribution of g on the earth's surface. Differentiating (3.14).

$$\frac{\partial U}{\partial l} = \frac{1}{2\pi} \iint g(x_0, y_0, 0) \frac{\partial}{\partial l} \frac{1}{r} \, dS \qquad (3.15)$$

In particular, if $g(x_0, y_0, 0)$ represents the vertical magnetizing force $Z(x_0, y_0, 0)$ and l the vertical direction,

$$-\frac{\partial U}{\partial z} = Z(x, y, z) = \frac{1}{2\pi} \iint + \frac{z}{r^3} Z(x_0, y_0, 0) \, dS \qquad (3.16)$$

Similarly the components of the magnetizing force in the x and y directions can be determined at any point (on or above the earth's surface) from the distribution of $Z(x_0, y_0, 0)$ alone. Clearly, therefore, horizontal magnetic intensity measurements on the surface of the earth will not yield any additional information if sufficiently dense and precise vertical intensity measurements are available (cf. p. 26).

The inherent indeterminancy of the inverse problem of gravity

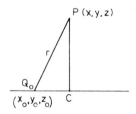

Fig. 25.

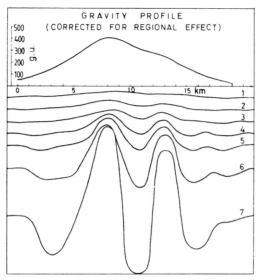

Fig. 26. Alternative basement reliefs (1–7) each of which explains a given gravity anomaly exactly ($\delta = 200 \text{ kg m}^{-3}$) [53].

and magnetic measurements made on the earth's surface will now be appreciated. The interpretation of such measurements in terms of subsurface mass distributions can never be unique unless some external control (e.g. geological information, drill-hole data, etc.) is available. The mere fact that the calculated anomaly of some mass distribution, whose geometrical parameters are appropriately adjusted, agrees everywhere with the measured values, is no guarantee that the distribution occurs in reality, however good the agreement may be.

Fig. 26 is an example where a given gravity anomaly has been explained exactly by each of the alternative basement reliefs. Automatic optimization techniques in no way reduce the fundamental ambiguity as the example in Fig. 27 shows. Either of the two complex structures (a) or (b) explains the measured gravity within the limits of accuracy. The structures have the finite lengths shown perpendicular to the plane of the figure, but do not extend symmetrically on either side of the plane. The dotted structures lie entirely on one side of the plane of the figure.

3.9.2 Analytic continuation

The procedure of calculating the values of a regular function in a certain domain D_2 when its values are given in D_1 where D_1 and

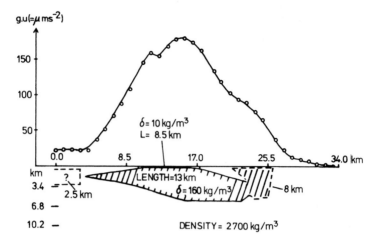

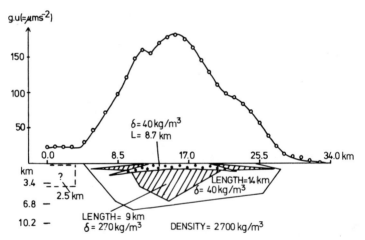

Fig. 27. Alternative structures producing the same gravity anomaly within the accuracy of measurements.

D_2 have a common boundary on which the two values are equal, is known as analytic continuation. Equation (3.14) is the fundamental integral of upward continuation of potential fields and (3.15) and (3.16) are its special cases. The problem of calculating $U(x, y, z)$ from a knowledge of $\partial U/\partial z(x_0, y_0, 0)$ is a straightfoward one of numerical integration. The faithfulness with which U can be reproduced at a higher level, that is farther from the masses, is only limited by the extent of the area on $z = 0$ within which $\partial U/\partial z$ is given, the denseness of the data and the fineness of the mesh (size of $dS = dx_0\, dy_0$) into which the datum plane can be divided for numerical integration. Transferring the origin to the plane $-z$, (3.16) becomes

$$\frac{\partial U}{\partial z}(x, y, 0) = \frac{z}{2\pi} \iint \frac{1}{r^3} \frac{\partial U}{\partial z}(x_0, y_0, z)\, dx_0\, dy_0 \qquad (3.16a)$$

which may be regarded as an integral equation for determining $g(x_0, y_0, z) = \partial U/\partial z(x_0, y_0, z)$ on a deeper level when $g(x, y, 0) = \partial U/\partial z(x, y, 0)$ is given on the surface. This process is called downward analytic continuation. It is a more difficult problem than upward continuation. Its formal solution, however, is easily obtained (Appendix 7) by using the convolution theorem.

The gravity field on a deeper level, calculated analytically starting from the field given on the surface, begins to fluctuate between positive and negative values at some stage. If the given field contains rapid variations of gravity (or magnetic intensity) with distance, indicating shallow sources, the fluctuations start to occur at a comparatively small depth. However, no matter how smooth the original field, the downward continued field will always start to fluctuate at *some* depth. This depth may in certain cases agree with the depth to the top of the masses producing the entire given field but strictly, it represents the maximum depth at which a density distribution of one sign only may lie and explain the original field [55]. However, the reader should note that this information can generally be obtained much more easily by other means (see next section) and the calculation of downward continued fields towards this end does not repay the labour involved (even using digital computers).

3.10 Depth determinations

If we postulate some regular shape for an anomalous mass, it is usually possible to devise rules which unambiguously determine the depth to its top surface or centre of gravity. As in magnetic interpretation (p. 38), the rules are often based on the distance at which the gravity anomaly or its horizontal gradient falls to a given fraction of the maximum value. Jung [56, 57] has derived a number of such rules and also given some geometrical constructions for rapid depth determinations. It is perhaps worth emphasizing that the converse is not true; that is, the geometry of a mass cannot be uniquely determined by assuming a depth to its top surface or centre of gravity. For, it can be shown that if, to any such mass capable of explaining a given gravity field, we add a distribution of the type $A \sin px$ with an arbitrarily great amplitude, the attraction of this extra distribution can be kept less than any quantity, however small, if the wavelength of the distribution is kept sufficiently short.

When no assumptions are made about the anomalous body the shallowest possible mass distribution may lie on the surface of the earth itself, or, if we refer to a volume distribution, it may lie with its top surface just touching the earth's surface at one or more points, unless geological or other data preclude such distributions. On the other hand it can be shown that the top surface of a distribution cannot lie at an arbitrarily *large* depth, but that there must be a limiting depth to it. For the rapid estimation of this maximum possible depth (h) from the observed gravity field, Smith [58, 59] has obtained a variety of important, rigorous results. Some of them are briefly discussed below. They apply to gravitating bodies whose density contrasts with the surrounding rock is either entirely positive or entirely negative. No other restrictions need be placed on the magnitude or variation of the density contrast or on the shape and location of the body. The assumption is thus very light and can be satisfied in a very large number of geological situations.

(1) If Δg_{max} and $\Delta g'_{max}$ are respectively the maximum values of the gravity anomaly and its horizontal gradient, then

$$h \leqslant 0.86 \, \Delta g_{max}/|\, \Delta g'_{max}\,| \qquad (3.17)$$

The inequality is particularly suitable when Δg_{max} and $\Delta g'_{max}$ occur close to each other.

(2) For all x we have

$$h \leqslant 1.50 \, \Delta g(x)/|\, \Delta g'(x)\,| \qquad (3.18)$$

This inequality can be used when only part of the whole anomaly is known.

For 'two-dimensional' density distributions such as those in Figs. 22b and c the numerical factor may be replaced by 0.65 in the inequality (3.17) and by 1.00 in (3.18).

(3) If $|\,\delta\,|_{max}$ is the absolute value of the maximum density contrast and $\Delta g''$ the second horizontal gradient of Δg,

$$h \leqslant 5.40 \, G \, |\,\delta\,|_{max}/|\, \Delta g''\,|_{max} \qquad (3.19)$$

This inequality may be inverted to find the lower limit of $|\,\delta\,|_{max}$ if a plausible value for h can be assigned on other data. If it is assumed that $\delta \geqslant 0$ throughout the body, the inequality can be improved by replacing the symbol $|\delta|_{max}$ by $1/2 |\delta|_{max}$.

The above formulas have been generalized by Smith [60] in an interesting manner. For instance, suppose that the body lies wholly between the planes $z = h$, $z = l$. Let $\overline{\Delta g}(d)$ be the average value of $\Delta g(x, y)$ around a circle of radius d lying in the plane $z = 0$ and having its centre at $(x, y, 0)$.

$$D = \overline{\Delta g}(d) - \Delta g(x, y, 0) \tag{3.20}$$

Then Smith has shown that

$$|D| \leqslant G |\delta|_{max} d[J(\alpha) - J(\beta)] \tag{3.21}$$

where $\alpha = h/d$, $\beta = l/d$, $J(t)$ is the function plotted in Fig. 28 and $|\delta|_{max}$ is, as before, the maximum density contrast. For large values of t, $J(t) = 1.752/t$ while $J(0) = 6.591$.

The inequality (3.21) can be used in a number of ways

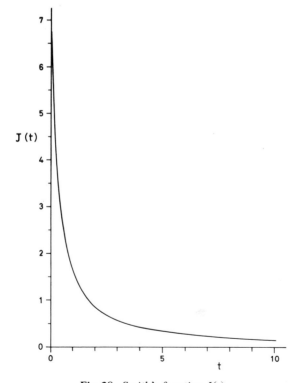

Fig. 28. Smith's function $J(t)$.

depending upon the information already available about the body (see, for example Section 3.13). If it is assumed that $\delta \geqslant 0$ throughout the body $|\delta|_{max}$ may be replaced by $1/2|\delta|_{max}$.

3.11 Determination of total mass

It is interesting to note that although gravity measurements alone cannot uniquely determine the distribution of anomalous masses, they do provide a unique estimate of the *total* anomalous mass. This is a corollary of Gauss' flux theorem. Let us describe a closed surface bounded by the hemisphere ABC (Fig. 29) and its circular intersection AC with the earth's surface. The total normal flux across the surface is $4\pi GM$ where M is the anomalous mass enclosed. As ABC becomes large, half the flux $(2\pi GM)$ passes across this surface of the hemisphere and the other half across the circle AC. From the definition of the normal flux we then obtain,

$$\iint\limits_{AC} \Delta g \, dS = 2\pi GM \tag{3.22}$$

where Δg is the gravity anomaly on AC. In practice, AC will comprise the area of measurements and the integration will be replaced by a summation. If Δg is expressed in g.u. and the element of area ΔS in square metres, we get from (3.22),

$$M = 2.39 \, \Sigma \, \Delta g \, \Delta S \text{ metric tons} \tag{3.22a}$$

If the body has a density ρ_1 and is embedded in a rock of density ρ_2, its *actual* mass will be

$$2.39 \, \frac{\rho_1}{\rho_1 - \rho_2} \, \Sigma \, \Delta g \, \Delta S \text{ tons} \tag{3.22b}$$

It will be seen that the anomalous mass, that is, the difference between the mass of a body and the mass of the country rock occupying an identical volume, can be determined without making any assumptions whatsoever except the physically trivial ones

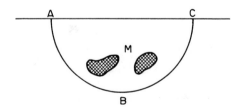

Fig. 29. Theorem of Gauss.

inherent in Gauss' theorem. However, the estimate of the actual mass requires an assumption concerning the *ratio* of ρ_2 and ρ_1.

In Equation (3.22) 2π may be replaced by an appropriately smaller solid angle if AC is not sufficiently large compared to the depth of the centre of gravity of M (cf. [61]).

3.12 Vertical derivatives of gravity

Let us suppose that we have two point masses at depths a_1 and $a_2 (> a_1)$. Their maximum anomalies Δg_1 and Δg_2 will be in the proportion a_2^2/a_1^2. The maximum values of $\partial \Delta g/\partial z$ will be in the proportion $a_2^3/a_1^3 (> a_2^2/a_1^2)$, those of $\partial^2 \Delta g/\partial z^2$ in the proportion $a_2^4/a_1^4 (> a_2^3/a_1^3)$ and so on. Thus, the successive vertical derivatives of gravity accentuate the relative effect of the shallower point mass. Now, the vertical derivative of gravity of any order, at any point on the earth's surface or above it, can be calculated merely from the knowledge of the distribution of gravity on the earth's surface. This follows from Equation (3.15) for we may perform repeated differentiations on $1/r$ with respect to z under the sign of integration and evaluate the resulting double integral numerically.

Many different methods of varying precisions have been proposed for computing the vertical derivatives from surface data [60]. We must remember, however, that a knowledge of the vertical derivatives does not in any way reduce the fundamental ambiguity in gravity interpretation. The reason is that the derivatives are directly deducible from the gravity field and therefore do not contain any additional information not inherent in the original data. This fact will not be altered even if the derivatives were directly measured by suitable instruments, alone or in addition to the gravity. It should be realized that derivative maps necessarily contain maxima and minima which have little structural significance and are only the result of straightforward algebraic properties.

Another property of the vertical derivatives is as follows. Suppose that we have two equal point masses at the same depth a. The gravity profile will then show two maxima when the masses are separated by a large distance (s) but a single maximum when they are close together. The vertical derivatives will present two principal maxima which also coalesce into one for some value of s. It can be shown that whereas the gravity maxima merge into one another when $s = a$, the principal maxima of the second derivative, for example, do not merge until a somewhat smaller value of s, namely $0.639\,a$, is reached. Thus, if the coalescence of the two maxima is chosen as the only criterion, the vertical derivatives of gravity would seem to have a greater resolving power than the gravity. There are, however, other criteria, e.g. the width of the gravity curve, which yield as much resolving power as the derivatives.

The above discussion concerns idealized cases. In practice, we generally have surface irregularities. Their effect on the vertical derivatives increases rapidly with the order of the derivative so that if, say, the second derivative is measured directly, the topographic corrections to it will be considerably more uncertain than the corresponding corrections to the gravity.

3.13 Illustrations of gravity surveys and interpretation

3.13.1 West Midlands (England)

This example is taken from a paper by Cook *et al.* [62]. The contours of the Bouguer anomalies in the area of measurement are shown in Fig. 30. The general geologic section is as follows. Rocks of Triassic age (Keuper Marl and Sandstones) are underlain by Coal Measures which contain coal seams and rest upon strata of Old Red Sandstone. Beneath the O.R.S. follow the rocks of Silurian and Cambrian age. All the strata are generally flat or have gentle dips but there is considerable faulting. Igneous intrusions occur at some places. The Trias is exposed in the Forest of Arden and in the area between the Forest of Wyre and South Staffordshire coalfields. In these coalfields and in the Warwickshire coalfield to the east the surface rocks are composed of Coal Measures.

The first step in the interpretation is to estimate the attraction of the surface rocks. Their thickness is fortunately well known over most of the area. However, this estimate cannot be made directly because although the density of the surface rocks may be known, the attraction is proportional to the difference of this density from that of the underlying rocks, which is not known. Cook therefore proceeds by plotting the anomalies against the thicknesses of the surface rocks concerned, say Coal Measures, at the observation stations. It will be appreciated from Section 3.7 (cf. Equation (3.9)) that the points thus obtained will lie approximately on a straight line with a slope 0.419×10^{-3} δ g.u./m $(0.128 \times 10^{-4}$ δ mgal/ft$)$ where δ is the difference of the density of the Coal Measures from that of the rocks below them. The plot for the Warwickshire coalfield (Fig. 31) yields

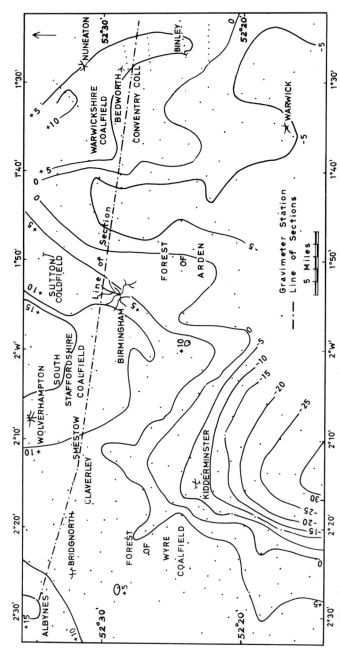

Fig. 30. Bouguer anomalies in West Midlands, England [62], contours in milligal, 1 mgal = 10 μms^{-2} = 10 g.u.

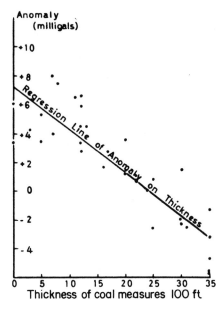

Fig. 31. Regression of anomaly on thickness [62]. (1 mgal = 10 g.u.).

$\delta = 230 \pm 20$ kg/m³. Similarly the density difference between the Trias and the underlying rocks (Coal Measures) is found to be 250 ± 140 kg/m³. The Bouguer anomalies then must be corrected for the attraction of the known Trias and Coal Measures, adopting the above density differences.

Fig. 32 shows the residual anomalies along a line from the Albynes to Bedworth. They indicate a large structure under the Trias cover between the Birmingham Fault and the Western Boundary Fault of the Warwickshire coalfield. It seems plausible from the geology that the low anomaly values in this part are caused by a slab of Coal Measures resting on the heavier Palaeozoic rocks. The western boundary of the slab is the known Birmingham Fault. Assuming the eastern one to be a vertical step the thickness of the slab can be estimated from Equation (3.10d) to be about 4000 ft. The maximum horizontal gradient over a vertical step is $2\pi G \delta \ln(D/d) = 48$ g.u./1000 ft. The observed value is only 10 g.u./1000 ft which indicates that the eastern margin is not vertical but slopes down (to the west) probably as a result of a number of step faults as shown.

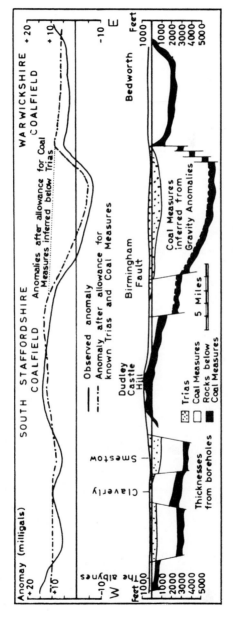

Fig. 32. Section along line in Fig. 30.

91

3.13.2 Camaguey Province, Cuba

The measurements discussed here were included in a prospecting campaign for chromite ore (Davis *et al.* [63]). The area is relatively flat except for a few scattered hills, mine dumps and excavations. The chromite deposits occur in serpentinized peridotite and dunite near their contact with felspathic and volcanic rocks. The Bouguer anomalies are shown in Fig. 33. The isoanomalous lines clearly indicate that apart from the local, roughly circular disturbance in the centre, there is a general ('regional') gravity trend consisting of an increase in the values from the southwest to the northeast (dotted contours). The residual anomalies obtained on subtracting this trend are shown in Fig. 34.

The maximum residual anomaly is 2.1 g.u. and its maximum horizontal gradient 1.6 g.u./20 m. Substituting in the Smith inequality (3.17) we get $h \leqslant 23$ m for the depth to the top surface of the chromite deposit. In fact, the ore was found to be exposed inconspicuously in a pit and later uncovered more fully by a shallow trench as shown.

The average value of the anomaly around a circle of radius

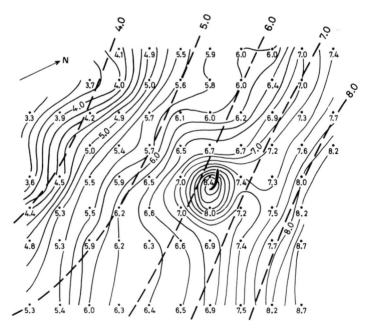

Fig. 33. Bouguer anomalies on a chromite deposit, values and contours in g.u. [63].

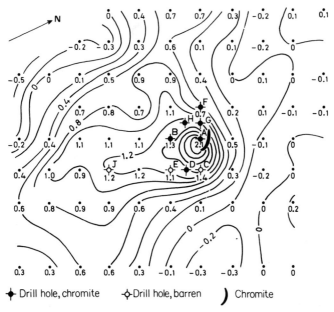

Fig. 34. Map in Fig. 33 after regional correction. Residual anomalies in g.u.

$d = 20$ m centred at the maximum is 0.98 g.u. Thus, in the inequality (3.20) $|D| = 2.10 - 0.98 = 1.12$ g.u. If we suppose that the ore body lies entirely between 0 and 60 m (the latter limit is suggested by the drilling results), $\alpha = 0$ and $\beta = 60/20 = 3.0$. From Fig. 28 we read off $J(0) = 6.591$ and $J(3.0) = 0.579$. Then assuming $\delta \geqslant 0$ throughout the body the inequality (3.21) becomes

$$\tfrac{1}{2}G\delta d \times (6.591 - 0.579) \geqslant 1.12 \times 10^{-6}$$

leading to a minimum density difference $\delta = 280$ kg/m³ between the chromite ore and the surrounding serpentinized peridotite. The latter is reported to have a density of 2500 kg/m³ so that the density of the chromite deposit must everywhere be equal to, or greater than, 2780 kg/m³. This, in turn, means that the grade of the deposit must be better than 13.9 per cent chromite (by volume) in all parts of it. Note that in these estimates we have not made any assumption as to the shape of the ore body, its depth, or its location sidewise.

Since the grid in Fig. 34 is a close and regular 20-m square array we may associate $\Delta S = 400$ m² with every observation point for estimating the anomalous mass according to (3.22). After inserting interpolated points between the last two rows in Fig. 34 we obtain $\Sigma\Delta g$ = 30.5 g.u. for the 86 points in the figure. The total anomalous mass is therefore 2.39 x 400 x 30.5 = 29 000 metric tons. If the body causing the anomaly has a density of, say, 4000 kg/m^3 and lies in serpentinized peridotite its actual mass will be 77 000 tons (cf. 3.22b). The ore ascertained from the drill holes amounts to only 24 000 tons. That the gravity anomaly is not entirely accounted for by the known chromite is evident also from the high anomaly values to the south of the maximum. The remaining mass, of course, need not be chromite.

3.14 Note on the Eötvös torsion balance

Before the development of modern gravimeters an instrument known as the Eötvös torsion balance was in considerable use for the detection of the local anomalies in the earth's gravitational field. Its principle, illustrated by Fig. 35, is similar to that of Cavendish's celebrated balance for the determination of G. In fact, the latter type of balance could also be used for gravity surveys.

In the Eötvös balance two masses m, m', each about 0.025 kg, are suspended at different levels by a platinum iridium wire. Suppose that an anomalous mass is situated below the instrument and on the reader's side. Its attraction on m will be greater than its attraction on m' so that a couple will act on the balance beam turning it out of the plane of the paper. An equilibrium will be reached when the couple is neutralized by the torsion moment of the suspension. By orientating the instrument in six (strictly speaking only five) different azimuths and observing the

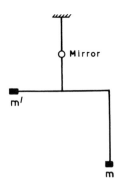

Fig. 35. Principle of the Eötvös torsion balance.

deflections in each case, the following quantities, where U is the gravitational potential, can be determined:

$$\frac{\partial^2 U}{\partial x^2} - \frac{\partial^2 U}{\partial y^2} \; , \; \frac{\partial^2 U}{\partial x \, \partial y} \; , \; \frac{\partial^2 U}{\partial x \, \partial z} \; , \; \frac{\partial^2 U}{\partial y \, \partial z}$$

The quantities $\partial^2 U / \partial x \, \partial z$ and $\partial^2 U / \partial y \, \partial z$ are the gradients of $\partial U / \partial z$, that is, of g, in the x and y directions. Adding them vectorially we get the direction and magnitude of the horizontal gradient of gravity at the observation point.

Another quantity that has frequently been used in the interpretation of torsion balance data is:

$$\left[\left(\frac{\partial^2 U}{\partial x^2} - \frac{\partial^2 U}{\partial y^2} \right)^2 + 4 \left(\frac{\partial^2 U}{\partial x \, \partial y} \right)^2 \right]^{1/2}$$

It can be shown that this quantity is equal to $g(1/R_1 - 1/R_2)$ where R_1 and R_2 are the principal radii of curvature of the equipotential surface at the observation point. $g(1/R_1 - 1/R_2)$, which has the same dimensions as the gradient of gravity, has been called the horizontal directive tendency (H.D.T.) or, by some writers, the curvature difference. To gain some idea of its nature, it may be noted that above a ridge of heavy rock, for example, the H.D.T. vectors will be parallel to the ridge. If the ridge is lighter than the surrounding rock the vectors will be perpendicular to the ridge.

The detailed theory of the Eötvös balance is complicated. For an excellent account of it the reader is referred to Eve and Keys [64].

A gradient of 10^{-9} newton/kg m $= 10^{-3}$ g.u./m($= 10^{-6}$ mgal/cm) is termed an Eötvös unit (E). The northward gravity gradient in latitude $45°$N, for instance, is 8.1 E (3.2). The torsion balance has a sensitivity of about 1 E and certainly ranks as one of the most sensitive physical instruments. Unfortunately, due to the fact that the instrument takes a relatively long time in reaching equilibrium and, due to the necessity of measuring the deflections in five or six azimuths, one complete observation takes several hours. A gravimeter observation on the other hand needs hardly more than 1–5 min. The torsion balance measurements require, moreover, elaborate corrections for the topographic irregularities in the vicinity of a station, including features such as houses, walls, ditches, etc. The modern gravimeters have therefore completely replaced the torsion balance in routine surveys.

3.15 Derivation of Formula (3.10c)

Provided the boundary of a closed area is piecewise continuous, an area integral can be converted into a line integral along the boundary, according to the relation:

$$\iint \frac{\partial u}{\partial z}\, dz\, dx = \oint u \cos (n, z)\, dl$$

where $u(x, z)$ is continuous, differentiable function, dl denotes an element of the boundary and $\cos(n, z)$ the direction cosine of the normal to it. This theorem due to Gauss will be found to be proved in any standard text on the calculus.

Choosing P as the origin in Fig. 22b the vertical component of the gravitational attraction at it due to an infinitesimal element $dx\, dz$ of the prism (density δ) is given by $2G\delta z\, dx\, dz/(x^2 + z^2)$ which is an elementary result (cf. (3.10b)). The total effect of the prism Δg is obtained by integrating this with respect to x and z. But the elemental attraction can be expressed as

$$G\delta dx\, dz\; \frac{\partial}{\partial z}\; \{\ln (x^2 + z^2)\}$$

so that by putting $u = \ln (x^2 + z^2)$ we get

$$\Delta g = G\delta \oint \ln (x^2 + z^2) \cos (n, z)\, dl$$

Now, the contour integral reduces simply to one along the two horizontal edges of the prism since along the vertical edges $\cos(n, z) = 0$. Thus

$$\Delta g = G\delta \cos(n, z) \int \ln(x^2 + z^2)\, dx$$

The integral here is elementary. Evaluating it by parts we get

$$\Delta g = G\delta \cos(n, z)[x \ln(x^2 + z^2) - 2x + 2z \tan^{-1}(x/z)]_{x_1}^{x_2}$$

where x_1, x_2 are the x-coordinates of the corners of the prism. Bearing in mind that $\cos(n, z) = -1$ along the upper face $(z = d)$ and $+1$ along the lower face $(z = D)$ we evaluate the bracketed expression, and shifting the origin to O in Fig. 22b, immediately get (3.10c).

The method can be extended to any prism whose cross-section is a right-angled polygon with sides parallel to the x- and z-axes. Let $1, 2, \ldots N$ be the corners (obviously even in number) of such a polygon, numbered anticlockwise, and let x_i, z_i be the coordinates of the corner i with P as the origin. Then the gravity anomaly at P is given by

$$\Delta g = 2G\delta \sum_{1}^{N} (-1)^i \{(x_i/2) \ln (x_i^2 + z_i^2) + z_i \tan^{-1} (x_i/z_i)\}$$

It is very easy to program this equation on a digital computer or a desk calculator. Kolbenheyer [65] has shown that the derivatives of the gravitational attraction of the prism can also be calculated by similar simple formulas and that the method can be extended to three-dimensional structures without difficulty.

4 Electrical methods

4.1 Introduction

The electrical properties of the sub-surface can be explored either electrically or electromagnetically. The two electrical methods to be discussed in this chapter are: (a) self-potential (SP) and (b) earth resistivity (ER). The induced polarization (IP) method, although it can be classified as an electrical method from the field-operational point of view, is discussed separately in the next chapter. The SP method was used by Fox as early as 1830 on the sulphide veins in a Cornish mine but the systematic use of the SP and ER methods dates from about 1920.

4.2 Self-potential

The SP method, as its name implies, is based upon measuring the natural potential differences which generally exist between any two points on the ground. These potentials, partly constant and partly fluctuating, are associated with electric currents in the ground. We shall revert to the fluctuating part in a later chapter. The constant and unidirectional potentials are set up due to electrochemical actions in the surface rocks or in bodies embedded in them.

Ranging normally from a fraction of a millivolt to a few tens of millivolts self-potentials sometimes attain values of the order of a few hundred millivolts and reveal the presence of a relatively strong sub-surface 'battery cell'. Such large potentials are observed as a rule only over sulphide and graphite ore bodies. Over sulphides at any rate they are invariably negative (Fig. 36). Similarly they are negative on graphite and graphitic shales [66] as well as on magnetite, galena and other electronically conducting minerals.

98

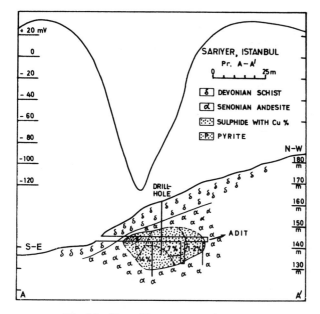

Fig. 36. SP profile across a pyrite mass.

4.2.1 Field procedure

The measurement of self-potentials is quite easy. Any millivolt-meter with a sufficiently high input impedance may be connected to two electrodes driven some 10–15 cm into the ground and read off. Alternatively a usual potentiometer circuit of the null type can also be employed with advantage. The electrodes must be non-polarizable (e.g. Cu in $CuSO_4$ solution or Pt-calomel in KCl solution); ordinary metal stakes will not generally do since electrochemical action at their contact with the ground tends to obscure the natural potentials.

Two alternative procedures are in use for SP surveys. The electrodes, say 20 m apart, are advanced together along staked lines or else one electrode connected to one end of a long cable on a reel is kept at a base point while the other electrode, the reel, and the voltmeter, are carried to different points as the cable is laid off. A new base point is chosen when the cable length is exhausted. The first procedure measures essentially the gradient of the potential.

The *formal* interpretation of SP anomalies is often carried out using formulae closely similar to those for the anomalies of magnetized bodies. It must be remembered, however, that SP anomalies are greatly affected by local geological and topographical conditions and these factors must be very carefully considered in SP work. For instance, it is often found that high ground is more negative than low ground, indicating that the electric current tends to flow uphill. An exceptionally high SP value (-1842 mV) associated with alunite (a basic sulphate of Al and K), but to a large extent due to topographic causes, has been reported from a mountain top near Hualgayoc, Peru [67].

4.2.2 Origin of self-potentials

There has been considerable speculation concerning the electrochemical mechanism producing self-potentials. The normal potentials observed over clays, marls and other sediments present no great problem. They can be explained by such well-known phenomena as ionic layers, electrofiltration, pH differences and electro-osmosis. It can be mentioned for instance that water flowing through the sand on a sea-beach sets up electrical potentials due to electrofiltration. Observations of electrofiltration or streaming potentials have been used in the USSR [68] to detect water leakage spots on the submerged slopes of earth-dam reservoirs. Concentration or pH differences can be used to map water-bearing zones of crushed or fractured rocks [66]. However, the large potentials observed over sulphide ores and graphite are more difficult to explain. Most theories have attributed the sulphide potentials to the oxidation of parts of the ore above the water table, but such an explanation cannot fit graphite which does not normally undergo significant oxidation.

The most detailed theory of the mechanism so far is that proposed by Sato and Mooney [69]. Since we are concerned with an electronic conductor (the ore) in contact with an ionic conductor (electrolytes in the country rock), there must be an exchange of electrons and ions at their boundary. The directions of current and ion flow implied by a negative centre over a sulphide are shown in Fig. 37. Evidently, electrons are being supplied to the upper end of the ore body, whereas oxidation of this end will require a liberation of electrons. Sato and Mooney conclude therefore that self-potentials cannot be caused directly by the near-surface oxidation of the ore. They propose instead that there is an oxidation potential difference (Eh) between the substances in solution above the water table and those below. The ore body, being a good electronic conductor, merely serves to transport the electrons from the reducing agents at depth to the

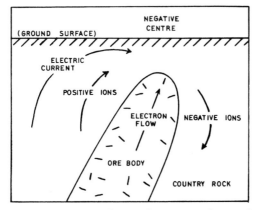

Fig. 37. Current and ion flow in the vicinity of an ore body.

oxidizing ones at the top, without itself directly participating in the electrochemical action.

Sato and Mooney are able to explain the order of magnitude of the self-potentials observed in the field with the above theory. The theoretical estimates are, however, definitely on the lower side. This fact and some others (e.g. the existence of large potentials even when the ore is totally submerged under the water table) indicate, however, that the problem of origin of sulphide potentials is not yet completely solved.

4.3 Earth resistivity

In the ER method a direct, commutated or low frequency alternating current is introduced into the ground by means of two electrodes (iron stakes or suitably laid out bare wire) connected to the terminals of a portable source of e.m.f. The resulting potential distribution on the ground, mapped by means of two probes (iron stakes or, preferably, non-polarizable electrodes), is capable of yielding the distribution of electric resistivity below the surface. The method has been used mainly in the search for water-bearing formations, in stratigraphic correlations in oil fields and in prospecting for conductive ore-bodies.

4.3.1 Potential distribution in a homogeneous earth

Consider a point electrode on the surface of a homogeneous isotropic earth extending to infinity in the downward direction and having a resistivity ρ. Describe a hemispherical shell of radius r and thickness dr around the electrode. If the current passing through the electrode into the ground is $+I$, the potential difference across the shell will be $dV = -I\rho\,dr/2\pi r^2$. Integrating, we get for the potential at a distance r from a point current source.

$$V(r) = \frac{I\rho}{2\pi}\frac{1}{r} \qquad (4.1)$$

The total potential at any point is $V = V(r) - V(r')$ where r' is the distance from the negative current electrode. It can be shown that in a homogeneous earth the fraction of the total current confined between the surface and the horizontal plane at a depth z is $(2/\pi)\tan^{-1}(z/L)$ where L is half the distance between the current electrodes.

We see from this that as much as 50 per cent of the total current in a homogeneous earth never penetrates below the depth $z = L$ and as much as 70.6 per cent never below $z = 2L$. The current will evidently penetrate deeper the greater the electrode separation.

The current penetration in a non-homogeneous earth composed of strata with horizontal interfaces has been worked out by Muskat and Evinger [70]. Fig. 38 is reproduced from their paper.

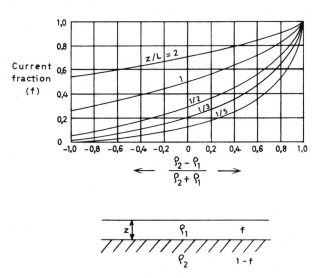

Fig. 38. Penetration of electric current in a two-layer earth.

4.3.2 Apparent resistivity

Let C_1, C_2 be the current electrodes, positive and negative respectively, and P_1, P_2 the potential probes. If ΔV is the potential difference between P_1 and P_2 it follows from (4.1) that

$$\rho = 2\pi \frac{\Delta V}{IG} \tag{4.2}$$

where

$$G = \frac{1}{C_1 P_1} - \frac{1}{C_2 P_1} - \frac{1}{C_1 P_2} + \frac{1}{C_2 P_2}$$

($\Delta V/I = R$ is of the nature of a resistance). It is advisable to use a null method in measuring ΔV in order to eliminate the influence of the contact resistances at P_1, P_2.

In an actual case, ρ will vary on altering the geometrical arrangement of the four electrodes or on moving them on the ground without altering their geometry. That is, R will not be directly proportional to G as on a homogeneous earth. The value of ρ, obtained on substituting the measured R and the appropriate G in (4.2), is called the *apparent resistivity* (ρ_a). It can be calculated for given electrode arrangements for a number of subsurface resistivity distributions.

The apparent resistivity is a formal, rather artificial concept and should not be construed, in general, as representing the average resistivity of the earth or any other similar magnitude. The artificiality will be evident from the fact that negative values of apparent resistivity are perfectly possible [71]. For a proper assessment of this quantity one must always bear in mind the configuration with which it has been obtained.

4.3.3 Electrode configurations

A variety of electrode arrangements are possible but we shall restrict ourselves to three very common linear arrays shown in Fig. 39. In the *Wenner* configuration the separations between the adjacent electrodes are equal (a) so that (4.2) reduces to

$$\rho_a = 2\pi a R \tag{4.3}$$

In the *Schlumberger* configuration $P_1 P_2 (=2l) \ll C_1 C_2 (=2L)$. At the centre of the system, Equation (4.2) can be written as,

$$\rho_a = \frac{\pi L^2}{2l} \frac{\Delta V}{I} = \frac{\pi L^2}{I} \left(-\frac{\mathrm{d}V}{\mathrm{d}r} \right) \tag{4.4}$$

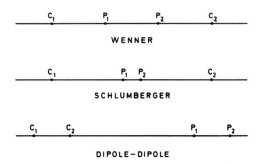

Fig. 39. Some electrode configurations.

where, obviously, dV/dr is the surface gradient of V, that is, the electric field at the centre. If $l \ll L/5$, Equation (4.4) gives a ρ-value correct to within a couple of per cent.

In the *dipole–dipole* array $P_1 P_2$ are outside $C_1 C_2$, each pair having a constant mutual separation a. If na is the distance between the two innermost electrodes (C_2, P_1) then Equation (4.2) gives

$$\rho_a = \pi n (n + 1)(n + 2)a \frac{\Delta V}{I}$$

Measurements are taken by increasing n in steps. If $n \gg 1$ each electrode pair may be treated as an electric dipole. Non-collinear dipole arrangements have also been used, especially in the USSR. The ρ_a values obtained with them bear simple relations to the value obtained (4.4) when the Schlumberger array is used [72].

The advantage of dipole–dipole arrays is that the distance between the current source and the voltage receiver systems can be increased almost indefinitely, being subject only to instrument sensitivity and noise, whereas the increase of electrode separations in the Wenner and Schlumberger arrays is limited by cable lengths.

It may be of interest to note that, by virtue of Helmholtz's reciprocity theorem in the theory of electric circuits, the ratio $\Delta V/I$ and hence ρ_a will be unaltered if the current and potential electrodes (in any configuration) are interchanged. This is true even if the earth is nonhomogeneous.

The properties of the sub-surface may be explored by two main procedures often called, by analogy, electric sounding (or drilling) and electric mapping (or trenching). In electric sounding with the Wenner arrangement the distance a is increased by steps, keeping the midpoint of the configuration fixed. If the Schlumberger configuration is used, only the current electrodes are moved outward symmetrically, keeping $P_1 P_2$ fixed at the centre.

In electric mapping with the Wenner method an electrode array with a fixed a is moved on the earth (usually along staked profiles) and the apparent resistivity is determined for each position. Alternatively, the two potential electrodes having a relatively small, constant separation in the Schlumberger configuration are moved between the fixed current electrodes to different positions. In this case, G in Equation (4.2) will vary with the position of the probes. This arrangement is also known as gradient array or mapping.

Since the fraction of the current penetrating to deeper levels increases with the electrode separation, electric sounding primarily provides information about the variation of electric conductivity with the depth. On the other hand, electric mapping measurements are influenced in the first place by lateral variations in the conductivity. They are therefore best suited for detecting local, relatively shallow non-homogeneities in the ground. The two procedures are thus complementary to a large extent.

4.3.4 Inverse problem of ER measurements

The problem of determining the distribution of the conductivity σ within the ground, when the surface potential due to a current electrode is given amounts (in isotropic media) to integrating the equation

$$\text{div}(\sigma \nabla V) = 0 \qquad (4.5)$$

It was shown by Slichter [73] and Langer [74] that if σ is a function of depth only the above equation possesses a unique solution. Then the sub-surface distribution of σ can be calculated from a knowledge of the surface potential produced by a single point electrode without any further physical or geological data.

In the general case when σ is not a function of z only, Stevenson [75] proved that Equation (4.5) possesses no unique solution, but that the problem can be made determinate by adding another 'dimension'. Thus, for example, if the surface potential is measured everywhere for all positions of a point electrode on some curve on the surface, the distribution of σ can, in principle, be uniquely determined.

It is instructive to contrast the present inverse problem with its theoretical possibilities of unique solutions with the inverse magnetic and gravity problems which are fundamentally indeterminate. It must be pointed out, however, that the determination of σ solves the *electric* problem but not necessarily the *geological* one, the reason being that given rock types and geological formations are not associated with definite resistivities except in a broad and general manner (see Section 4.9).

4.4 Layered earth

Stevenson's theorem is largely of theoretical interest. In practice one must, as a rule, resort to a trial-and-error technique in which the surface potential due to a current source is calculated for an assumed conductivity distribution and compared with the observations.

One type of conductivity distribution that adequately describes many geological situations is that represented by an earth composed of several horizontal strata. This model is of particular importance in prospecting for ground water by earth resistivity methods.

In the simplest case we have a horizontal layer of thickness h_1 and resistivity ρ_1 overlying a second homogeneous medium of resistivity ρ_2 (Fig. 40). The potential of a point electrode C, through which a current I is passing into such an earth, was first calculated by Hummel [76, 77] by the method of electric images. The potential is given by the sum of (i) the potential of C in a

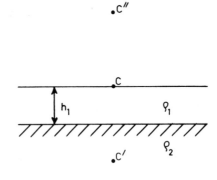

Fig. 40.

semi-infinite medium of resistivity ρ_1, that is $I\rho_1/2\pi CP$ according to Equation (4.1) and (ii) the potentials of fictitious current sources C', C'', C''', ..., etc. of appropriate strengths, where C' is the image of C in the plane $z = h$, C'' is the image of C' in $z = 0$ and so on ad infinitum. We then get for the potential at a surface point P:

$$V(r) = \frac{I\rho_1}{2\pi} \frac{1}{r} \left[1 + \sum_{n=1}^{\infty} \frac{2k^n r}{(r^2 + 4n^2 h^2)^{1/2}} \right] \tag{4.6}$$

where $r = CP$ and $k = (\rho_2 - \rho_1)/(\rho_2 + \rho_1)$.

The potential can also be expressed as

$$V(r) = \frac{I\rho_1}{2\pi} \frac{1}{r} \left[1 + 2r \int_0^{\infty} K(\lambda, k, h) J_0(\lambda r) d\lambda \right] \tag{4.7}$$

where $K(\lambda) = k \exp(-2\lambda h)/[1 - k \exp(-2\lambda h)]$ and J_0 is the Bessel function of order zero as shown in Appendix 6, a form that is indispensable for treating the general case of an arbitrary number of layers.

With the knowledge of V it is possible to calculate ρ_a from Equation (4.2) for any electrode configuration. We shall confine ourselves to the Schlumberger array, Equation (4.4). The ratio ρ_a/ρ_1 in this case is shown in Fig. 41 on a double logarithmic plot as a function of L/h_1, for values of ρ_2/ρ_1 from 0 (perfectly conducting sub-stratum) to ∞ (perfectly insulating sub-stratum).

It will be seen that ρ_a approaches ρ_1 when the current electrode separation is small compared to the thickness of the top layer and ρ_2 when it is large. The transition from ρ_1 to ρ_2 is, however, smooth and no simple general rule based on specific properties of the curve (e.g. the gradient) can be devised to find the thickness h_1.

With the addition of a third layer (h_2, ρ_2) sandwiched between the top layer (h_1, ρ_1) and the sub-stratum (ρ_3), the problem becomes much more complicated. The apparent resistivity curve can then take four basic shapes known as Q (or DH, descending Hummel), A (ascending), K (or DA, displaced anisotropic) and H (Hummel type with minimum), depending upon the relative magnitudes of ρ_1, ρ_2, ρ_3 (Fig. 42). In every case, however, ρ_a approaches ρ_1 for small values of L and ρ_3 for large ones. At intermediate values of L it is influenced by the resistivity of the middle layer.

Master curves such as those in Fig. 41 cannot be conveniently

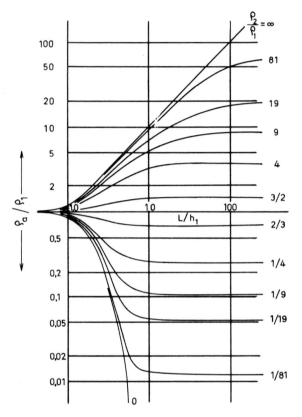

Fig. 41. Master curves of Schlumberger apparent resistivity on a two-layer earth.

presented for the three-layer case in a single diagram. The examples in Fig. 42 will nevertheless give an idea of the type of curves to be expected. Extensive catalogues have been published elsewhere [78, 79] and can be used with advantage in interpreting the field data.

The manner of using them or Fig. 41, provided that the data can be plausibly referred to the number of layers for which the master curves are constructed, is very simple. The observed ρ_a is plotted against $L = C_1 C_2 /2$ on a transparent double logarithmic paper with the same modulus as the master curves paper. Keeping the respective axes parallel, the transparent paper is slid on various master curves in succession until a satisfactory match is obtained with some curve (if necessary an interpolated one). The value of $C_1 C_2 /2$ coinciding with the point 1.0 on the x-axis of the

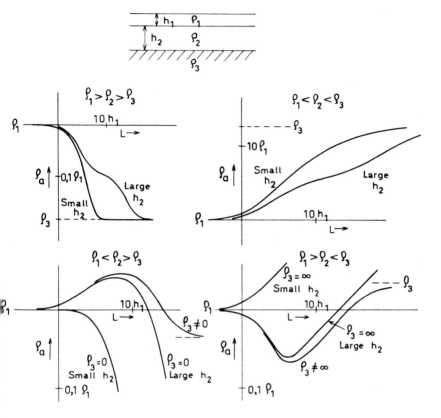

Fig. 42. Q (or DH), A, K (or DA) and H-type curves in vertical electrical sounding (VES).

matching master curve gives h_1 and the value of ρ_a coinciding with the point 1.0 on the y-axis gives ρ_1. The values $h_2, \rho_2 \ldots$ etc. are obtained from the appropriate parameters belonging to the matching master curve. The h and ρ values thus obtained are, however, subject to the principle of equivalence (p. 119).

The validity of the above procedure is easily grasped when one notes that either side of (4.6) or (4.7), for example, can be expressed in a dimensionless form. It is noted that there is again no simple relation between the coordinates of the turning points or the inflection points of the ρ_a curves and the parameters h_1, h_2 and ρ_2.

The curve matching technique is practicable only when the number of layers is small, say up to 4. Even for four layers the number of reasonable parameter combinations is so large and the collection of curves so bulky that interpretation by matching becomes generally impracticable and for a larger number of layers it is virtually impossible. Fortunately modern developments described later have rendered the curve matching technique largely obsolete.

4.5 Kernel function and resistivity transform

The potential distribution on an earth composed of an arbitrary number of layers is again given by the same Equation (4.7) but with $K(\lambda)$ a function of all strata thicknesses and resistivities. $K(\lambda)$ is known as the kernel function of resistivity. For a three-layer earth, for example,

$$K(\lambda) = \frac{k_1 u_1 + k_2 u_1 u_2}{1 - k_1 u_1 - k_2 u_1 u_2 + k_1 k_2 u_2} \tag{4.8}$$

where $u_1 = \exp(-2\lambda h_1)$, $u_2 = \exp(-2\lambda h_2)$ and h_1, h_2 are the thicknesses of the two layers (resistivities ρ_1, ρ_2) overlying an infinite sub-stratum (the third 'layer', resistivity ρ_3) while k_1, k_2 are the resistivity-contrast coefficients $(\rho_2 - \rho_1)/(\rho_2 + \rho_1)$ and $(\rho_3 - \rho_2)/(\rho_3 + \rho_2)$ respectively.

In passing it should be noted that in integral equation theory it is $J_0(\lambda r)$ that is known as the kernel function. It is therefore preferable to refer to $K(\lambda)$ as the characteristic function rather than as the kernel function, but the latter usage is widespread in geophysical literature.

From (4.4) and (4.7) we get, for the Schlumberger array,

$$\rho_a = \rho_1 \left(1 + 2r^2 \int_0^\infty K(\lambda) J_1(\lambda r) \lambda \, d\lambda \right) \tag{4.9}$$

since $J_0'(x) = - J_1(x)$ where J_1 is the Bessel function of order 1. Note that in obtaining (4.9) from (4.7) the potential gradient must be doubled to include the effect of the negative current electrode of the array also. In (4.9) r is then half the current electrode separation.

Using the following limit of a Lipschitz integral in Bessel function theory

$$\lim_{c \to 0} r^2 \int_0^\infty e^{-\lambda c} J_1(\lambda r) \lambda \, d\lambda = 1$$

Equation (4.9) becomes

$$\rho_a(r) = r^2 \int_0^\infty T(\lambda) J_1(\lambda r) \lambda \, d\lambda \qquad (4.10)$$

where $T(\lambda) = \rho_1\{1 + 2K(\lambda)\}$.

$T(\lambda)$ is known as the resistivity transform and is easily written down for an arbitrary number of layers by means of recurrence formulae (Appendix 6). Note that whereas $K(\lambda)$ is dimensionless, $T(\lambda)$ as defined above has the dimensions of ohm metre.

Now, according to Hankel's transformation in Bessel function theory, if we have a function such that

$$f(r) = \int_0^\infty F(\lambda) J_n(\lambda r) \lambda \, d\lambda$$

then

$$F(\lambda) = \int_0^\infty f(r) J_n(\lambda r) r \, dr$$

Applying the transformation to (4.10) we have

$$T(\lambda) = \int_0^\infty \frac{\rho_a}{r} J_1(\lambda r) \, dr \qquad (4.11)$$

This equation shows that the resistivity transform can be computed unambiguously from a measured ρ_a curve by numerical integration and implicitly contains all the information about the layered earth. It plays an important role in the modern theory of the interpretation of resistivity sounding data as will be seen below.

4.6 Determination of layered earth parameters

The electric conductivity of a layered earth is a function of the depth only. According to the Slichter–Langer theorem, therefore, the knowledge of the surface potential due to a point electrode should suffice to determine the thicknesses and resistivities of the various layers. Two interpretation methods, the auxiliary point and the Pekeris–Kofoed method, based on this approach are available. A number of other methods embodying essentially a trial-and-error or optimization approach (e.g. Ghosh, Inman and Johansen methods) have also been proposed in recent years. These latter type of methods are numerically (and geologically) the more efficient ones.

4.6.1 The auxiliary point method

This method can be traced back to Hummel although it has been subsequently refined and improved by several workers [80]. Let us take a three-layer case (ρ_1, ρ_2, ρ_3) of Fig. 42 with $h_2 \gg h_1$. Clearly, as long as the current electrode separation does not exceed a certain value the apparent resistivity curve will not differ appreciably from a two-layer case with the same ρ_2/ρ_1 as that in the three-layer case under consideration. At large electrode separations the third layer, that is, the infinite sub-stratum (ρ_3) will influence the measurements.

Hummel showed that for sufficiently large separations the apparent resistivity curve obtained is virtually the same as for a two-layer case with the same sub-stratum but with an overlying layer of thickness $H = h_1 + h_2$ and a resistivity ρ_m given by $H/\rho_m = h_1/\rho_1 + h_2/\rho_2$. It will be noticed that ρ_m is derived by applying Kirchhoff's law for resistances in parallel. The equation is easily rearranged as

$$\frac{\rho_m}{\rho_1} = \frac{\rho_2/\rho_1}{\rho_2/\rho_1 - 1 + H/h_1} \frac{H}{h_1} \qquad (4.12)$$

By matching the initial branch of a measured resistivity curve with an appropriate curve in Fig. 41 we obtain ρ_2 and h_1 (ρ_1 is known from the asymptote of ρ_a for very small electrode separations). Similarly, by matching the branch obtained with large electrode separations we get ρ_3/ρ_m and H again with the help of Fig. 41. Since ρ_2/ρ_1 and H/h_1 are now known, ρ_m can be obtained from (4.12) and hence ρ_3, since ρ_3/ρ_m has been determined. The calculation for ρ_m is avoided if the right-hand side of (4.12) is plotted on an auxiliary master diagram as a function of the dimensionless parameter H/h_1. A family of curves for various ρ_2/ρ_1 values is obtained and ρ_m can be read off as the ordinate of the appropriate curve.

The method can be extended in principle to a sounding curve on any number of layers by the alternate use of the two-layer master curves in Fig. 41 and the family of auxiliary curves based on Equation (4.12). The 'drill' with the auxiliary point method can be illustrated by the hypothetical example in Fig. 43.

An inspection of the curve in the figure suggests that there must be at least four layers of which the second must be more resistive than the surface layer, the third must be less resistive than the second (as the descending branch shows) and the fourth (sub-stratum) must be more resistive than the third.

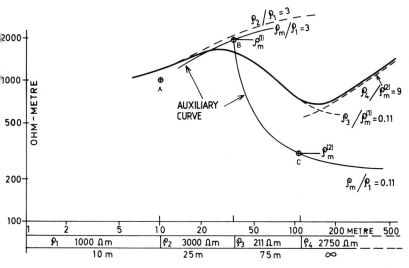

Fig. 43. Hypothetical example of auxiliary point method of VES interpretation.

A transparent paper on which the sounding curve has been traced is slid on Fig. 41 (the logarithmic modulus must be equal on both drawings) keeping the respective axes parallel to each other until a reasonably long portion of the first branch of the measured curve coincides with one of the master curves. In the example, the coinciding master curve is one for which $\rho_2/\rho_1 = 3$ (dashed in Fig. 43). The origin $(1, 1)$ of the master collection is marked on the tracing (circle A with cross in Fig. 43). The coordinates of A give ρ_1, h_1.

The tracing is now placed on the auxiliary diagram on which the family (4.12) has been drawn (again on a logarithmic scale of the same modulus) with A coinciding with the origin of the auxiliary diagram (the auxiliary point), the respective axes being parallel. The auxiliary curve for which $\rho_2/\rho_1 = 3$ is copied on the tracing for a sufficient length. This is the curve marked $\rho_m/\rho_1 = 3$ in Fig. 43.

The tracing is now again transferred to Fig. 41 and slid over it, keeping the axes parallel, *and with the origin always on the copied curve* $\rho_m/\rho_1 = 3$, until a further reasonably long portion of the measured sounding curve (in this example the descending branch) coincides with one of the curves in Fig. 41. In Fig. 43 this is the dashed curve $\rho_3/\rho_m^{(1)} = 0.11$. The origin of Fig. 41 is again marked on the tracing (point B). The coordinates of B give $h_1 + h_2$ and $\rho_m^{(1)}$ which is ρ_m of Equation (4.12).

The tracing is now once more placed on the auxiliary diagram and the curve for 0.11 copied. On retransferring the tracing to Fig. 41 and repeating the procedure above to obtain a coinciding master curve ($\rho_4/\rho_m^{(2)} = 9$) we locate the point C, the coordinates of which give $\rho_m^{(2)}$ and $h_1 + h_2 + h_3$ where

$$\frac{h_1 + h_2 + h_3}{\rho_m^{(2)}} = \frac{h_1}{\rho_1} + \frac{h_2}{\rho_2} + \frac{h_3}{\rho_3}$$

The resistivities of the various layers are easily obtained by multiplying the ratios found by ρ_1, $\rho_m^{(1)}$ and $\rho_m^{(2)}$. The complete interpretation is shown at the bottom in Fig. 43.

The most serious limitation of the auxiliary point method is that it requires the thickness of each successive layer to be much greater than the *combined thicknesses* of all the overlying layers. The method has nevertheless been extensively used and can give satisfactory results in experienced hands, especially if certain modifications of Equation (4.12) are used [80, 81]. The risk of misleading results is, however, very great since the basic condition is a severe geological restriction. Now that fast computer methods are available the main use of the auxiliary point method is in locating a trial solution for subsequent optimization.

4.6.2 Pekeris' direct method

This method of direct interpretation due to Pekeris [82] is based on Slichter's analysis. We shall merely consider the practical procedure; the theoretical justification will be found in Pekeris' original paper.

It is easy to measure the surface potential $V(r)$ of a single point electrode by removing the other current electrode to great distance. We compute the function

$$k(\lambda) = \lambda \int_0^\infty V(r) J_0(\lambda r) r \, dr \qquad (4.13)$$

by numerical integration and plot $\ln |f_1(\lambda)|$ against λ where

$$f_1(\lambda) = \frac{k(\lambda) + 1}{k(\lambda) - 1}$$

For large λ the points will lie on a straight line with a slope $2h_1$ and an intercept $\ln(1/k_1)$ with the y-axis where $k_1 = (\rho_2 - \rho_1)/$

$(\rho_2 + \rho_1)$. If all the points lie on a straight line $h_2 = \infty$ (two-layer case); otherwise with h_1 and k_1 thus determined we proceed to calculate another closely similar function $f_2(\lambda)$, the plot of whose logarithm gives in turn h_2 and $k_2 = (\rho_3 - \rho_2)/(\rho_3 + \rho_2)$.

The process is continued by calculating a third function f_3 if all the points do not lie on a straight line, that is, if $h_3 \neq \infty$ and so on.

This method implies no restrictions on the relative magnitudes of h_1, h_2, h_3 . . . and is thus quite general. Unfortunately it requires considerable computation. However, Koefoed has recently developed the method into a rapid practical procedure [83]. He starts by constructing $K(\lambda)$ from the observed ρ_a curve (4.9) by means of a small number of standard curves. Next he defines a modified kernel function $G_n(\lambda) = K(\lambda)/\{1 + K(\lambda)\}$ and this he treats in essentially the same manner as $f(\lambda)$ above to obtain successive strata parameters.

The main task in the direct method is the rapid and accurate calculation of the kernel function (or the resistivity transform) from measured data. Towards this end Patella [84] has devised a fast method of calculating $T(\lambda)$ in (4.11) that is suitable for pocket-size electronic calculators. His subsequent procedure for determining layer parameters is essentially the same as Kofoed's.

4.6.3 Ghosh's method
In this trial-and-error method the entire ρ_a curve of a model layered structure is calculated for comparison with the measured curve. The calculations are simple enough to be carried out on a pocket-size programmable calculator and the method has the advantage over the direct procedures of the previous section that the measured data are not tampered with.

In essence Ghosh evaluates the right-hand side of Equation (4.10), but with the help of a set of coefficients (Table 7), taking advantage of the fact that $T(\lambda)$ in (4.10) is an algebraic expression involving no more complicated functions than the exponential. For the mathematical justification the reader should consult Ghosh's original paper [85] but the practical procedure, divided into two stages, is simply as follows.

(a) In the first stage, sampled values T_m ($m = 0, 1, 2, \ldots$) of $T(\lambda)$ are calculated from the recurrence relations of Appendix 6, for successive values of λ in the ratio $10^{1/3}$, that is, $2.154.1/\lambda$ is the distance along the x-axis and the sampling rate corresponds on a logarithmic plot to $3T_m$-values per decade. Any starting value of λ

Table 7   Ghosh coefficients for calculating ρ_a (Schlumberger array)

b_{-3}	b_{-2}	b_{-1}	b_0	b_1	b_2	b_3	b_4	b_5
0.0225	−0.0499	0.1064	0.1854	1.9720	−1.5716	0.4018	−0.0814	0.0148

may be used but it is convenient to start with $\lambda = 1$ and calculate for $\lambda < 1$ as well as >1.

(b) In the second stage, the sample values of apparent resistivity are obtained as

$$(\rho_a)_m = \sum_{j=-3}^{5} b_j T_{m-j} \quad (m = 0, 1, 2, \ldots) \tag{4.14}$$

where b_j are the 9 coefficients in Table 7. Due to the particular manner in which the b_j's have been calculated by Ghosh, each $(\rho_a)_m$ value obtained from Equation (4.14) refers not to the abscissa of the corresponding T_m but to an abscissa that is 5% to the left. Thus, for example, $(\rho_a)_m$ corresponding to T_m at $1/\lambda = 21.54$ must be plotted at $1/\lambda = 20.47$ etc. On a logarithmic plot of modulus 62.5 mm this simply means shifting the whole calculated curve by $62.5 \log(0.95) = -1.4$ mm.

If desired, a second set of transform values may be defined in between those of the first set and the calculation in (4.14) repeated to obtain ρ_m values at a closer spacing.

It can be seen from the form of Equation (4.14) that in order to obtain $(\rho_a)_m$ values within a particular range of current electrode separation it is necessary to have 5 extra T_m samples to the left of the range and 3 to the right. The accuracy in the calculated ρ_a values is of the order of 3 per cent.

If a large computer is available it is not necessary to restrict the number of coefficients to 9 or the number of sampled $T(\lambda)$ values to 3 per decade. Thus, Johansen [86] used 140 fixed coefficients, calculated to 8 decimals, and sampled $T(\lambda)$ at 10 points per decade, achieving a relative accuracy of the order of 1 part in 10^{-6} in the computation of ρ_a, the time needed being less than one second on a CDC 6400 computer, even for a model containing as many as 10 layers.

In Ghosh's method, and Johansen's modification of it, the trial-and-error procedure can be tedious and it is desirable to let the computer seek the optimum model by an iterative procedure. We come thus to the methods of the next section.

4.6.4 Optimization methods

A number of automatic methods of electric sounding interpretation have been proposed but only two of them, one due to
Inman *et al.* [87] and the other due to Johansen [88] will be
briefly touched upon here. In contrast to many other optimization
methods these two methods leave the measured data intact and
use them only for comparison with a purely theoretical model, a
simpler and safer approach than transforming the data.

Let $P_j(j = 1, 2, \ldots, m)$ be the n resistivities and $n - 1$ layer-
thickness ($m = 2n - 1$).

Let y_i be the measured values of ρ_a for the separations between
two current-electrodes x_i, $i = 1, 2, \ldots, n$. The problem is to
minimize the sum of squares

$$S = \sum_{i=1}^{n} \{y_i - \rho_a(x_i, P_j)\}^2 \tag{4.15}$$

with respect to P_j where $\rho_a(x_i, P_j)$ is the model value given by
Equation (4.10). ρ_a and P_j are non-linearly connected. Equation
(4.10) is linearized by expanding ρ_a in a Taylor series around a
starting model P_j^0 and discarding all terms of higher order than the
first. This gives

$$S = \sum_{i=1}^{n} \left\{ y_i - \rho_a(x_i, P_j^0) - \sum_{j=1}^{m} \frac{\partial \rho_a}{\partial P_j} \delta P_j \right\}^2 \tag{4.16}$$

where the derivatives are evaluated at the point P_j^0.

The minimization of (4.16) determines the corrections δP_j. The
process is repeated by setting $P_j = P_j^0 + \delta P_j$ until there is no further
decrease in S. The P_j values in this case give the desired optimum
parameters. Inman *et al.* work with (4.15) calculating ρ_a and the
derivatives by using infinite series.

Johansen replaces y_i and ρ_a by $\ln y_i$ and $\ln \rho_a$ respectively, thus
linearizing the problem to a great extent to start with and
calculates ρ_a and the derivatives by Ghosh-type coefficients. This
approach leads to a considerable decrease in the time necessary to
obtain the optimum solution. The overall time for Johansen's
method is about 10 seconds on a CDC 6400.

In both approaches the optimization procedure can be pro-
grammed to produce certain extremal parameter sets showing the
limits within which some or all of the parameters can stay and yet
give a satisfactory agreement with measurements. This is a very

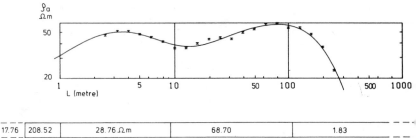

| 17.76 | 208.52 | 28.76 Ω m | 68.70 | 1.83 |

Fig. 44. A VES curve and its interpretation (Area: Wilhelmsborg, Denmark).

valuable feature of the procedures suggested by Inman *et al.* and Johansen, as will be seen below in the discussion of the principles of equivalence and suppression. Fig. 44 shows an example of a measured sounding curve and the layered structure deduced from it.

Another example of sounding is shown in Fig. 45 together with the geological section [89]. The object of this investigation was to locate permeable sand beds bearing fresh water. The interpretation

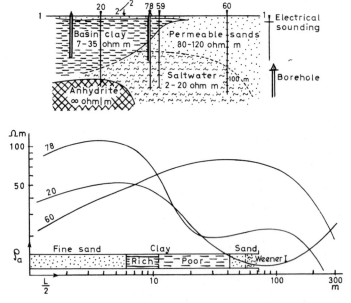

Fig. 45. VES curves and interpretation [89]. (L/2 in this figure has the same significance as *L* in Section 4.3).

of these graphs, which to a large extent led to the geological picture shown, is not easy although certain qualitative correlations will be readily recognized.

4.6.5 Principles of equivalence and suppression

In the actual application of the various interpretation methods to a particular field problem limitations are set by the maximum distance from the current source to which the electric field is given, and by the irregularities in the field due to surface non-homogeneities. Furthermore all measurements have a finite accuracy. On account of all these causes, widely different resistivity distributions may lead to ρ_a curves which, although they are not identical, cannot be distinguished in practice. This introduces ambiguity in the interpretation.

The mathematical formulation of two simple types of equivalence is easily obtained. Consider, for example, a relatively thin layer sandwiched between two layers whose resistivities are much higher than that of the sandwiched layer. The current flow in the earth will then tend to concentrate into the middle layer and will be almost parallel to the layer (Fig. 46a). The resistance of an elementary block of length Δl and cross-section $h\Delta m$ to such a current flow is $R = \rho\Delta l/(h\Delta m)$ and this will be unaltered if we increase ρ but at the same time increase h in the same proportion. Thus all such middle layers for which the *ratio* h/ρ is the same (within certain limits on h and ρ) are electrically equivalent.

On the other hand, if the resistivity of the middle layer is much larger than that of the layers on either side, the electric current will tend to avoid it and take the shortest route to the lower layer. The lines of current flow will be almost perpendicular to the layer (Fig. 46b). The resistance of an elementary block to this flow will be $R = \rho h/\Delta A$ where ΔA is the cross-section. In this case all layers for which the *product* $h\rho$ is the same are electrically equivalent so that, again, h and ρ cannot be determined separately.

If a thickness of a layer is very small compared to its depth (and

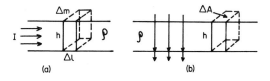

Fig. 46. Principle of equivalence.

its resistivity is finite) its effect on the ρ_a curve is so small that the presence of the layer will be suppressed.

In a general sequence of several layers most of the h and ρ parameters are subject to the principles of equivalence and suppression but the parameters can, in general, only vary within certain limits. These limits can be found by the optimization procedures discussed in the previous section. Table 8 shows some equivalences for the example in Fig. 44.

Almost perfect equivalence, when h and ρ can vary within extremely wide limits, occurs when a layer is either a very good or a very poor conductor in relation to its surroundings. The electrical determination of the depths to layers lying below a highly equivalent layer becomes most ambiguous unless either the resistivity or the thickness of the layer can be estimated from extraneous data (e.g. a trial borehole).

4.6.6. Note on Dar Zarrouk curves

Consider a prism of unit cross-section and vertical lateral faces cut out of a layered medium with n layers. Its electrical resistance to a current moving perpendicular to the layering will be

$$T_n = \sum_1^n h_j \rho_j \quad \text{('transverse unit resistance')} \quad (4.17)$$

Similarly we can define

$$S_n = \sum_1^n (h_j/\rho_j) \quad \text{('longitudinal unit conductance')} \quad (4.18)$$

The total thickness is

$$H_n = \sum_1^n h_j \quad (4.19)$$

Defining $\rho_\perp = T_n/H_n$ and $\rho_\parallel = H_n/S_n$ we see that the prism behaves as a whole like an anisotropic medium with a coefficient of anisotropy (or rather pseudo-anisotropy)

$$\lambda_n = (\rho_\perp/\rho_\parallel)^{1/2}$$

It can be shown that a layer of anisotropy λ and thickness H is indistinguishable electrically from an isotropic layer of a thickness $H\lambda$ and a resistivity $(\rho_\perp\rho_\parallel)^{1/2}$. Therefore the pseudo-anisotropic block is equivalent to an isotropic layer with the parameters

$$(\rho_m)_n = (T_n/S_n)^{1/2}, \quad H \text{ (equivalent)} = (T_n S_n)^{1/2} \quad (4.20)$$

Table 8 Some equivalences for the example in Fig. 44 (h in m, ρ in Ωm)

Solution of Fig. 44

	ρ_1	h_1	ρ_2	h_2	ρ_3	h_3	ρ_4	h_4	ρ_5
	17.76	0.48	208.52	0.52	28.76	8.26	68.70	99.50	1.83

Each of the following solutions fits the measured curve as well as the solution above. They form a subset of many possible equivalences obtained by extremizing various parameters of the problem.

1. *Models extremizing ρ_1*

h_1, ρ_1 may vary arbitrarily provided ρ_1 lies between extremes cited and h_1/ρ_1 lies between the corresponding limits (almost constant).

	ρ_1	h_1	ρ_2	h_2	ρ_3	h_3	ρ_4	h_4	ρ_5
ρ_1 max.	35.24	0.98	194.33	0.52	28.76	8.24	68.70	99.46	1.85
ρ_1 min.	6.06	0.17	442.04	0.26	28.98	8.45	68.74	99.27	1.90

2. *Models extremizing h_2*

h_2, ρ_2 may vary arbitrarily provided h_2 lies between extremes cited and $h_2\rho_2$ lies between the corresponding limits (almost constant).

	ρ_1	h_1	ρ_2	h_2	ρ_3	h_3	ρ_4	h_4	ρ_5
h_2 max.	17.66	0.46	105.61	1.07	28.35	8.00	68.58	99.80	1.84
h_2 min.	9.29	0.27	658.45	0.16	29.24	8.60	68.81	9.07	1.91

3. *Models extremizing h_4*

h_4, ρ_4 may vary arbitrarily provided h_4 lies between extremes cited and $h_4\rho_4$ lies between the corresponding limits.

	ρ_1	h_1	ρ_2	h_2	ρ_3	h_3	ρ_4	h_4	ρ_5
h_4 max.	17.28	0.48	218.08	0.52	27.56	7.29	66.09	109.78	0.62
h_4 min.	16.73	0.45	220.58	0.48	29.79	9.27	71.40	91.63	3.14

Suppose that this prism rests on an $(n + 1)$th block (thickness h_{n+1}, resistivity ρ) of identical cross-section. Then, considering only a sub-block of thickness h we can define two functions as extensions of (4.17) and (4.18).

$$T = T_n + h\rho$$
$$S = S_n + (h/\rho) \quad 0 \leqslant h \leqslant h_{n+1} \tag{4.21}$$

with the corresponding parameters

$$\rho_m = (T/S)^{1/2}, \quad H_{eq} = (TS)^{1/2} \tag{4.22}$$

The parametric equations (4.22) in conjunction with (4.21) define a function $\rho_m (H_{eq})$ in the interval $(T_n S_n)^{1/2} \leqslant H_{eq} \leqslant (T_{n+1} S_{n+1})^{1/2}$. The graph of $\rho_m (H_{eq})$ against H_{eq} is known as the Dar Zarrouk curve for that layer.

Starting from the first layer, for which the curve reduces to $\rho_m = \rho_1$, a Dar Zarrouk curve (or rather segment) can be constructed for each successive layer, either from the equations above or by the rapid method suggested by Orellana [90]. The Dar Zarrouk segments of two adjacent layers cut each other at an angle and are not smooth continuations of each other.

The theoretical interest of Dar Zarrouk curves lies in the fact, proved by Maillet [91], that the function $T = T(S)$ is sufficient to determine the potential on the ground surface and therefore the sounding curve. Consequently, it is not surprising that the aggregate of Dar Zarrouk segments for a given layered structure resembles the sounding curve on it.

With the advent of fast computer programs, and interpretation methods that can be run on pocket-size calculators, the importance of Dar Zarrouk curves, like that of the auxiliary point method, has naturally declined.

4.7 Vertical and dipping discontinuities

The case of two homogeneous rock formations separated by a plane vertical boundary (Fig. 47) can be treated by the method of images [e.g. 66] or, more generally by solving Laplace's Equation [92]. If we have a single point electrode C, the other electrode being at infinity, we get for the apparent resistivity defined by Equation (4.4).

$$\rho_a = \rho_1 \left[1 - k_1 \frac{r^2}{(2a - r)^2} \right] \quad r < a \text{ (medium 1)}$$
$$\rho_a = \rho_1 (1 + k_1) \qquad\qquad r > a \text{ (medium 2)} \tag{4.23}$$

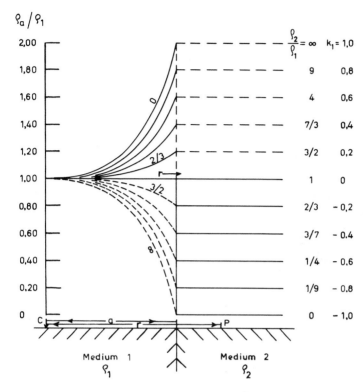

Fig. 47. Calculated Schlumberger ρ_a curves across a vertical contact.

with k_1 as in the last section. Typical curves are given in Fig. 47.

It will be seen that the apparent resistivity is discontinuous at the boundary. The discontinuity will be evident in practice as a steep gradient of the resistivity curve (Fig. 48).

In Fig. 49 are shown the apparent resistivity curves in another commonly encountered case, namely a thin conducting dike or vein cutting through the surrounding rock.

Figs. 47 and 49 incorporate two features which are worth a special mention since they apply to all work with the resistivity methods. Firstly, small resistivity contrasts cause comparatively large departures of the ratio ρ_a/ρ_1 from 1.0. Secondly, the ratio is practically unchanged whether the resistivity contrast is moderately large, say $\rho_2/\rho_1 = 10$ or very large, say $\rho_2/\rho_1 = 10\,000$. This is often known as the 'saturation effect'. Consequently, while the resistivity method is efficient in detecting small variations in the conductivity of the ground, it is ill-adapted for distinguishing, say,

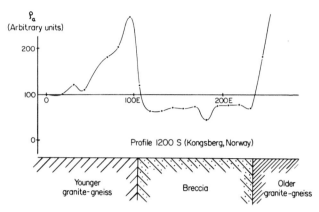

Fig. 48. Field example of ρ_a measurements across steep geological contacts [92].

a good conductor from a very good conductor, although both will be readily detected.

Dipping discontinuities were discussed by Maeda [93] and by de Gery and Kunetz [94] as far back as 1955. However, a reasonably practical attempt for the quantitative interpretation of sounding data on such discontinuities seems to be due to Lee [95]. In fact, Lee treats the case of a general two-dimensional surface separating two media and not only plane discontinuities.

Lee's method for determining the configuration of such a surface S (e.g. an undulating bedrock surface below a homogeneous top-layer) is briefly as follows. Measure ρ_a (in the sense of Schlumberger) on a profile in the strike direction z, say. Then it can be shown that for large values of λ

$$T(\lambda) = \int_0^\infty \frac{\rho_a}{z} J_1(\lambda z)\, dz \simeq \rho_1 \{1 + 2kN \exp(-2\lambda d)\}$$

$$= H(\lambda), \text{ say} \qquad (4.23)$$

where N is the number of times a horizontal cylinder of the smallest radius d, and axis along the profile of measurement, is tangential to S.

The parameters $2kN$ and $2d$ are easily found from a plot of $\ln|H/\rho_1 - 1|$ against λ for several large values of λ. Circles of radii d centred on the respective profiles are drawn in a plane representing the vertical plane perpendicular to the strike direction.

For a plane discontinuity $N = 1$. For undulating surfaces N for

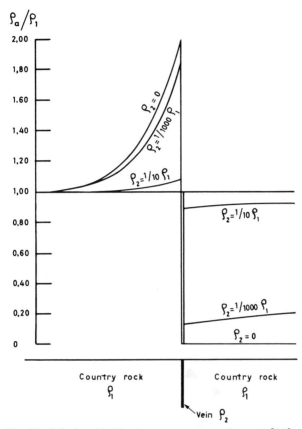

Fig. 49. Calculated Schlumberger ρ_a across a thin vein [92].

each profile must be determined by inspection of the various kN values in such a way that $k = (\rho_2 - \rho_1)/(\rho_2 + \rho_1)$ is the same for all profiles. Appropriately many lines tangential to each of the circles are finally drawn and the picture smoothed to obtain the simplest structure.

A degenerate case arises when S is a plane discontinuity perpendicular to the ground surface as in Fig. 47. For this $2k$ in (4.23) must be replaced by k.

4.8 Electrical mapping, anisotropic earth and logging

4.8.1 Mapping

The example in Fig. 48 demonstrates essentially the procedure that was termed electrical trenching or mapping in Section 4.3. In general, any lateral inhomogeneity in the conductivity manifests itself as a discontinuity or a more or less sharp gradient in the apparent resistivity curve or its distance derivative. Sometimes the curve takes a form which it is not easy to visualize intuitively. An interesting example is shown in Fig. 50a. Here the theoretical curve has been calculated for the Wenner arrangement taken across a hemispherical sink embedded in an otherwise homogeneous ground. Fig. 50b is a field curve over a deposit which approximates to this form [96].

In areas with a well-defined geological strike, more or less constant in direction, it is often advantageous to use line instead of point sources for electrical mapping. The potential of an infinitely long line electrode is given by $(I\rho/\pi)\ln r$ where I is the current per unit length. For a finite electrode the logarithmic term is somewhat more complicated [66].

Fig. 51 shows the results of an electrical mapping survey in an ore-bearing area using parallel electrodes, 200 m long and 1200 m apart. The potential probes (40 m apart) were moved along lines perpendicular to the current electrodes and between them. The country rock in the region, with apparent resistivities less than 1000 ohm m, is more or less uniformly impregnated with pyrite and pyrrhotite with mineable concentrations of chalcopyrite in places. The boundaries between the barren and the impregnated rock are clearly indicated by the steep gradient in the resistivity contours.

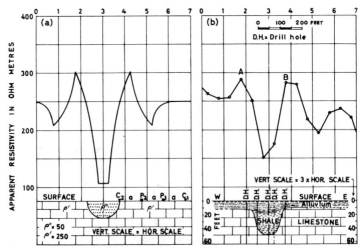

Fig. 50. Wenner ρ_a across hemispherical sinks [96].

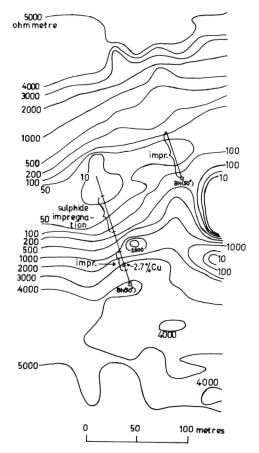

Fig. 51. Resistivity map of an area in North Sweden.

The electrical conductivity distributions considered in this chapter are rather simple cases. More general and complicated conditions have been treated by Alfano [97] and Vozoff [98] in several interesting papers.

4.8.2 Anisotropic earth

Although we have considered homogeneous as well as non-homogeneous earth it has been tacitly assumed that all the media concerned are electrically isotropic. However, many rocks such as shales, slates, laminated ores and even some bedrock formations are markedly anisotropic and serious errors can arise in the interpretation if this fact is neglected.

We must distinguish between two kinds of anisotropy: a micro-anisotropy in which the individual grains of a rock are anisotropic owing to their structure and a macro-anisotropy arising when a given formation contains several different facies (themselves isotropic or anisotropic) alternating in a regular fashion. In general the two effects will be superimposed. The main consequences of anisotropy bearing on the earlier discussions in this chapter may be stated as follows.

Let a homogeneous but anisotropic earth have resistivities ρ_l and ρ_t, parallel and transverse to its surface. Then the apparent resistivity ρ_a determined by a linear array of four point electrodes on the surface is $\sqrt{(\rho_l \rho_t)}$ while the apparent resistivity perpendicular thereto measured in a vertical borehole will be paradoxically enough, $\rho_l (< \sqrt{(\rho_t \rho_l)})$.

Secondly, any anisotropic layer of thickness h can be replaced by an isotropic layer of thickness $h\sqrt{(\rho_t/\rho_l)}$ and resistivity $\sqrt{(\rho_t \rho_l)}$ without altering the surface potential distribution of a point electrode. This evidently introduces an indeterminacy in the interpretation of resistivity data.

The ratio $\sqrt{(\rho_t/\rho_l)}$ is known as the coefficient of anisotropy. For layered rocks such as shales, gneisses etc. it is usually of the order 1.10–1.30. A more detailed description of anisotropy will be found in [80].

4.8.3 Electrical logging

Apparent resistivity determinations can be made by sinking an electrode array in a borehole. In practice one current electrode is often placed on the surface at a large distance from the borehole while the other current electrode and the two potential probes, with fixed mutual distances, are lowered in the hole, the electrodes being pressed against the walls by springs. A similar technique is also used for measuring self-potentials in boreholes.

These procedures of electric measurements in boreholes are known as electrical logging or coring. They have been used chiefly in oil fields for stratigraphic correlations from one borehole to another, it being found that one and the same geological formation within a restricted area is almost always associated with characteristic resistivity values or self-potentials.

The SP values in sedimentary formations are almost entirely concentration potentials or potentials arising due to diffusion of ions. It has been claimed [99] that weak formation-potentials can be enhanced by the addition of certain chemicals to the drilling mud. The subject of electric logging has been treated in great detail by Lynch [100].

4.9 The resistivity of rocks and minerals

The magnitude of the electric anomalies on a non-homogeneous earth depends upon the resistivity differences between different rocks, or, more correctly, upon factors of the type $(\rho_2 - \rho_1)/(\rho_2 + \rho_1)$. The resistivity of rocks is an extremely variable property ranging from about 10^{-6} ohm m for minerals such as graphite to more than 10^{12} ohm m for dry quartizitic rocks. Most rocks and minerals are insulators in the dry state. In nature they almost always hold some interstitial water with dissolved salts and therefore acquire an ionic conductivity which then depends upon the moisture content, the nature of the electrolytes and their concentration. The form of the pores in a rock plays a subordinate role in determining the conductivity.

Some minerals, notably graphite, pyrrhotite, pyrite, chalcopyrite, galena and magnetite (a ferrite) are relatively good (electronic) conductors (Table 9). A dissemination of such minerals within a rock can also make it a better conductor. Others, such as zinc blende, are also electronic conductors but very poor such at ordinary temperatures.

Practically all rocks and minerals are semiconductors, their resistivity decreasing with increasing temperature. Especially in the sulphide minerals, donor and acceptor impurities play a great part in determining the absolute magnitude of the electrical conductivity. The activation energies are however poorly known for most of the naturally occurring sulphides and oxides. Shuey [101]

Table 9 Electric resistivities (ohm m)

Rocks and sediments		Ores	
Limestone (marble)	$>10^{12}$	Pyrrhotite	$10^{-5}-10^{-3}$
Quartz	$>10^{10}$	Chalcopyrite	$10^{-4}-10^{-1}$
Rock salt	10^6-10^7	Graphite shales	$10^{-3}-10^1$
Granite	$5000-10^6$	Pyrite	$10^{-4}-10^1$
Sandstones	$35-4000$	Magnetite	$10^{-2}-10^1$
Moraine	$8-4000$	Haematite (†)	$10^{-1}-10^2$
Limestones	$120-400$	Galena	$10^{-2}-300$
Clays	$1-120$	Zinc blende	$>10^4$

†Stoichiometric Fe_2O_3 insulator.

has compiled the available data and discussed the subject from a quantum mechanical point of view.

The resistivity of porous, water-bearing rocks (free of clay minerals) follows Archie's law, $\rho = \rho_0 f^{-m} s^{-n}$ where ρ_0 is the resistivity of the water filling the pores, f is the porosity (volume fraction pores), s is the fraction of pore space filled by the water and n, m are certain parameters. The value of n is usually close to 2.0 if more than about 30 per cent pore space is water-filled but can be much greater for lesser water contents.

The value of m depends upon the degree of cementation or, as this is often well correlated with geologic age, upon the geologic age of the rock. It varies from about 1.3 for loose, Tertiary sediments to about 1.95 for well-cemented Palaezoic ones, but can be outside this range for individual formations [102].

5 Induced polarization

5.1 Introduction

If an electric current in the ground is interrupted the voltage across P_1, P_2 (Fig. 39) does not drop to zero instantaneously. It is found instead to relax for several seconds (or minutes) starting from an 'initial' value which is a small fraction of the voltage (V) that existed when the current was flowing (Fig. 52). This phenomenon has been termed *induced polarization* and is easily observed when electronically conducting minerals or clay minerals are present in the ground. It was noted in geophysical work by C. Schlumberger some time before 1920 but modern application of the phenomenon to geophysical exploration dates from about 1948 although several, largely inconclusive, experiments were reported between 1920 and 1948.

The phenomenon has been known to electrochemists studying the passage of electric currents in electrode—electrolyte systems and has in this connection been called *overvoltage*. That a very similar effect exists in pure dielectrics has also been known for a considerable time [103].

It takes a finite, although short, time before the voltage V is reached when a current is switched on. This implies that for an uninterrupted current flow induced polarization should manifest

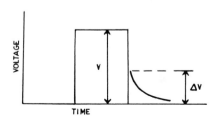

Fig. 52. IP phenomenon.

131

itself as a dependence of the ground impedance on the frequency of the current. This is, in fact, borne out by observations.

Thus, the IP phenomenon can be observed in time as well as frequency domain. It is an exceedingly complex phenomenon although it superficially resembles the discharge of a capacitor (time domain) or the variation of the impedance of an RC circuit (frequency domain). An adequate explanation of all aspects of the IP phenomenon has yet to be given but we shall consider some of the theories advanced later on.

5.2 Measures of IP

5.2.1 Time domain

When measurements are made by passing d.c. pulses of duration T in the ground the magnitude of the observed IP is often expressed as $\Delta V/V$ (millivolt per volt or per cent) where ΔV is the voltage remaining at a definite time t after current cut-off. This measure of IP, which we shall denote by P_t^T, is called the *polarizability*. Commonly used values of T are in the range of $1-20$ s while t is a fraction of T.

It is usual to send the d.c. pulse first in one direction and then in the reverse direction, after a current-off time following the measurement of P_t^T. The cut-off time is generally of the same order of magnitude as T. It is essential to choose t properly. It must be long enough for the electromagnetic induction effects in the ground to have substantially disappeared but short enough for ΔV not to fall below the sensitivity of the detecting device.

The ratio $\Delta V/V$ is independent of V, at least for the current densities in the ground in normal field operations. Sometimes the normalized time integral $(1/V)\int_{t_1}^{t_2}\Delta V_{\mathrm{IP}}\,\mathrm{d}t$ representing the area under the decay curve is used to express IP as millivolt second per volt or as millisecond. This measure is known as *chargeability* (denoted M_{t_1,t_2}^T).

A quantity that implicitly contains some information about the shape of the decay curve is the ratio L/M where M is as defined above and L is the corresponding quantity representing the area 'above' the curve, that is, between the dashed line in Fig. 52 and the decay curve between t_1 and another time t_3.

5.2.2 Frequency domain

In the frequency domain, the apparent resistivity of the ground is measured by any one of the innumerable electrode configurations possible (Section 4.3) at two frequencies F and $f(<F)$. IP is expressed as the apparent frequency effect $FE_{F,f} = (\rho_{af} - \rho_{aF})/\rho_{af}$. The term per cent frequency effect is also used if this ratio is expressed as a percentage change in ρ_a. f is usually in the range $0.05-0.5$ Hz and F in the range $1-10$ Hz.

Another frequency domain measure of IP is the *metal factor* $MF_{F,f} = A(\rho_{af} - \rho_{aF})/(\rho_{af}\rho_{aF}) = A(\sigma_{aF} - \sigma_{af})$ where A is a suitable numerical coefficient (10^5 or $2\pi \times 10^5$ are values that are often used) and the σ_a's are the apparent conductivities. In SI MF has the dimensions siemen per m.

The time domain measure corresponding to $MF_{F,f}$ will be IP/ρ_a where $IP = P_t^T$ or M_{t_1,t_2}^T, but this does not appear to have been used.

A third measure of the IP effect in frequency domain is the phase difference ϕ between the voltage between P_1, P_2 and the current injected into the ground. Maximum values of ϕ in practice are usually a few hundredths to a tenth of a radian at a frequency of, say, 1 Hz.

5.3 Origin of IP

5.3.1 Electrode and membrane polarizations

The electric conduction paths in the ground are normally ionic but they may be sometimes hindered to a greater or lesser extent by mineral particles (e.g. pyrite grains) in which the carriers of current are electrons.

It is well known that when a current passes across a metal electrode (electronic conductor) dipped in an electrolyte, charge can pile up continuously at the interface when all the processes in the electrolytic reaction are not equally rapid. This produces the familiar back e.m.f. or electrode polarization. The extra pile-up charge diffuses back into the electrolyte when the current is stopped, re-establishing the original equilibrium in which a thin layer of negative ions is fixed to the metal electrode. The IP observed over sulphide ores or other electronically conducting minerals like graphite or magnetite, is basically a manifestation of such 'electrode polarization'. The effect will be enhanced if the mineral grains are dispersed rather than in a compact mass since it is essentially a surface phenomenon and the polarization charge will be large owing to the large total surface of the particles. Values of more than 10 per cent for $\Delta V/V$, for example, are not uncommon on such ores (Fig. 97).

Induced polarization is also observed in the absence of electronically conducting minerals. The presence of clay particles appears to be a necessary condition for this effect for it is not observed on clean quartz sand or similar media devoid of clay minerals. 'Membrane polarization', as this effect has been termed, is most probably due to ionic exchanges and the setting up of diffusion potentials somewhat as follows.

The surface of clay particles, the edges of layered and fibrous materials or cleavage faces normally have unbalanced negative charges that attract a cloud of positive ions from the surrounding electrolyte. When an electric current is forced through a clay—electrolyte system, positive ions can readily pass through this cloud but negative ions are blocked forming zones of ion concentration. The return of the ions to the former equilibrium distribution after the current is stopped constitutes a residual current and appears as the IP effect. Superficially, as far as observations are concerned, membrane and electrode IP effects resemble each other. No diagnostic feature that can distinguish these two effects unambiguously has yet been found for field observations.

5.3.2 Macroscopic theories

It is beyond the scope of this monograph to discuss in detail the various attempts that have been made to explain the IP phenomenon quantitatively. For an excellent résumé of these the reader is referred to the monograph of Bertin and Loeb [104] and for a qualitative account to Sumner's monograph [105]. Here we shall indicate some of the main lines in these theories.

The two principal features of the phenomenon that must be explained by any IP theory are the shape of the decay curve and its counterpart in the frequency domain, namely the variation of the (complex) resistivity with frequency. It should be realized that the decay curve is not a simple exponential but more like a sum of several exponentials. In some cases the variation is also found to be of the type $\Delta V = A t^{-n}$ where t is the time after current cut-off.

Many of the experimental results including the above two can be conveniently approximated by the assumption that the current density j in the ground is given by

$$j = (\sigma + i\epsilon_{\mathrm{IP}}\omega)E \quad (i = \sqrt{(-1)}) \tag{5.1}$$

where σ and E are the conductivity and the electric field, ω is the angular frequency and ϵ_{IP} is a parameter of the nature of a dielectric constant.

The dielectric constant of rocks is of the order of $5-100 \epsilon_0$ where ϵ_0 is the dielectric constant of vacuum (8.854×10^{-12} farad per metre). Various estimates of ϵ_{IP} have been made from IP observations [104]. Even the lowest of these indicate $\epsilon_{IP} \approx 10^4 - 10^5 \epsilon_0$ and some of the highest ones $\epsilon_{IP} \approx 10^{11} \epsilon_0$! It is clear that ordinary dielectric properties cannot explain IP and the observed decay constants of IP voltage imply 'abnormally' high interface capacitances at mineral grains.

The macroscopic or phenomenological theories attempt to explain the observations by means of certain *ad hoc* parameters and relations, of which the above discussion gives one example.

Another such attempt is Seigel's assumption that the source of IP signal is a secondary current density vector $\mathbf{m} = m\mathbf{j_0}$ where m is a dimensionless parameter ('chargeability') and $\mathbf{j_0}$ is the current density at the end of a charging process. The electric potential in the medium is then given by the analogue of Equation (2.7) and Seigel showed [106] that the assumption amounts to an effective decrease of σ to $\sigma(1 - m)$ during the charging or the discharging processes. With these assumptions it is possible to calculate the IP response of polarizable bodies. This approach does not, of course, intend to explain the time dependence of IP.

In considering time dependence we recall that the voltage across a charged capacitor in parallel with a resistor decays exponentially with time. As mentioned above the IP decay curve is of a more complicated type. Various attempts have therefore been made to explain the observed time dependence of IP and the frequency dependence of the complex resistivity by simulating suitable, more elaborate resistor–capacitor combinations [104]. While such networks explain the results of specific experiments they cannot be said to explain the *physical* origin of IP.

5.3.3 Microscopic theories

The object of microscopic or physical theories is to explain the IP phenomenon by consideration of the forces acting on the electric charge carriers involved and their motion under these forces. Suppose, for example, that the ions in the ground electrolytes move only under diffusion forces obeying the diffusion equation

$$\frac{\partial C}{\partial t} = D \frac{\partial^2 C}{\partial x^2} \tag{5.2}$$

where C is the ion concentration (mole per m^3) and D the diffusion coefficient (m^2 s^{-1}), a suggestion made by Warburg already in 1899 for explaining the response of non-polarizable electrodes to alternating current.

If C undergoes a sinusoidal variation of frequency f at $x = 0$ the general solution of Equation (5.2) is known to be of the form $C = C_0 \exp\{-x(\pi f/D)^{1/2}\} \sin\{2\pi ft - x(\pi f/D)^{1/2}\}$ which suggests that the amplitude of the IP voltage should depend on $f^{1/2}$. Accurate experiments show that it in fact depends on f^α where $\alpha \neq \frac{1}{2}$. In the time domain it can be shown [104] that the time constants derived from this theory are several orders of magnitude smaller than those observed in practice for the IP phenomenon.

Simple diffusion is thus not sufficient to explain the IP phenomenon. Of the attempts made to improve upon the diffusion hypothesis two will be mentioned here.

Bertin and Loeb in the monograph cited above assume that the anions and cations in the electrolyte under consideration are subject to electrical forces in addition to diffusion forces. They then arrive at the following modification of (5.2):

$$\frac{\partial C}{\partial t} = D \frac{\partial^2 C}{\partial x^2} + uC \frac{\partial E}{\partial x} \tag{5.3}$$

(E = electric field V m^{-1}, u = ionic mobility m^2/V s). Equation (5.3) is assumed to be satisfied separately by the anions and cations.

Considering the situation at the end of the charging process when no flow of matter takes place any longer, Bertin and Loeb are able to explain the order of magnitude of ϵ_{IP} on the basis of (5.3).

Nilsson [107] starts with slightly different assumptions and taking into account the thermal agitation forces derives a non-linear differential equation for the electric field at a distance x from a plane electrode,

$$\frac{\partial E}{\partial t} = D \left(\frac{\partial^2 E}{\partial x^2} - \frac{ze}{kT} E \frac{\partial E}{\partial x} \right) \tag{5.4}$$

where z denotes the number of electrons lost per atom in the formation of the cations, e is the electronic charge, k is Boltzmann's constant and T is the absolute temperature.

A numerical integration of this equation under appropriate boundary conditions seems to reproduce the time dependence of

IP. Nilsson also makes the interesting suggestion that IP and SP have the same origin, the latter being the static field obtained as a function of x for $\partial E/\partial t = 0$. (In consulting Nilsson's paper the reader should note that it contains an inordinate number of misprints.)

It is interesting to note that if the concentration of ions is postulated to be proportional to the electric field Equation (5.3) is essentially the same as (5.4). The full implication of these equations and of other similar attempts is still to be worked out and the physical theory of IP is not yet in a final shape.

5.4 Electromagnetic coupling

Electromagnetic induction in the ground creates potential differences that are superimposed on those due to the true IP effect. These coupling effects can be troublesome in IP measurements when the current-supply lines are long and the ground is highly conductive. They must be assessed before the true IP effect can be estimated.

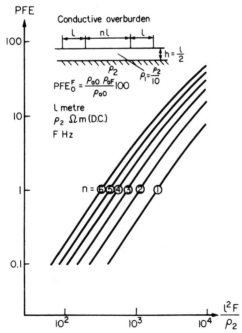

Fig. 53. Per cent frequency effect (PFE) due to electromagnetic coupling in frequency domain IP for a dipole–dipole configuration on the two-layer earth shown [104].

A universally applicable method of correcting for electromagnetic coupling does not exist. In practice one must assume an electrical structure for an area (e.g. a layered sequence) and calculate, for the electrode configuration in question, the spurious IP parameter magnitude that can arise due to electromagnetic induction. This is then subtracted from the measured value of the parameter.

Hohmann [108] among others has treated in some detail the problem of electromagnetic coupling in frequency domain methods. Figs. 53 and 54, after him, give an idea of the spurious per cent frequency effect that will arise for a dipole–dipole configuration (Section 4.3) on the two-layer earth shown. Similar calculations for the time domain method have been published recently by Rathor [109] and for the phase difference IP method by Wynn [110].

As a rule it is advisable to select a measurement technique that minimizes EM coupling effects rather than try to estimate them. For example for a Schlumberger array (Fig. 39) L for time domain work should be such that $\mu L^2/\rho \ll \sim 1$ s where μ is the magnetic permeability ($= 4\pi \times 10^{-7}$ Ω s/m for most practical purposes) and $\rho(\Omega$ m) the resistivity of the ground. For frequency domain work

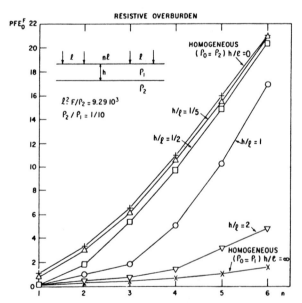

Fig. 54. Per cent frequency (PFE) due to e.m. coupling as a function of overburden thickness and n in a dipole–dipole array [104].

APPARENT RESISTIVITY (ρ_a) AND METAL FACTOR (MF)$_a$ ON
ZONE OF CONCENTRATED MINERALIZATION
LAKE CHIBOUGAMAU, QUEBEC

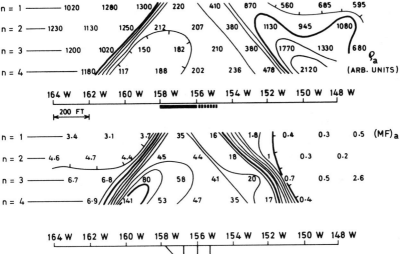

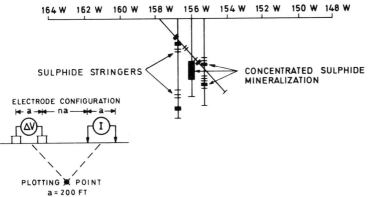

Fig. 55. A frequency domain IP survey with the dipole–dipole configuration [112].

Sumner [105] suggests that the ratio a/d (a = electrode spread and $d = \{\rho/\pi f\mu)\}^{1/2}$) should be less than 0.1.

Similarly, configurations in which the current electrodes line is at right angles to the potential electrodes line minimize the effect of electromagnetic coupling.

5.5 Example of an IP survey

Although IP measurements have sometimes been undertaken as vertical soundings [111] like the resistivity work described in the previous chapter, most of the present IP work is of the mapping type.

Fig. 55 is an example of a ρ_a and IP (frequency domain) survey [112]. It should be evident from Section 5.2 that an IP measurement necessarily involves a concomitant resistivity measurement. The measurements in Fig. 55 were made by using a number of dipole–dipole arrays (Fig. 39) with different n values. The manner of plotting the readings should be clear from the inset. A diagram obtained in this manner is often qualitatively regarded as representing 'an electrical vertical section' or rather 'pseudo-section' through the profile, since increasing n values give information from increasingly larger depths, but the plot must not be taken too literally. Vertical tabular bodies of constant thickness, for example, can be shown to give wedge-shaped anomalies on this plot. Thus the patterns in Fig. 55 do not by any means necessarily indicate a broadening of the mineralization towards the deeper levels.

An example of a time domain IP profile will be found in Fig. 97.

6 Electromagnetic continuous wave, transient-field and telluric methods

6.1 Introduction

If a time-varying electromagnetic field is produced on the surface of the ground currents will flow in sub-surface conductors in accordance with the laws of electromagnetic induction. These currents give rise to secondary electromagnetic fields which distort the primary field observed at any point on the surface. In general, the resultant field, which may be picked up by a suitable search coil, will differ from the primary field in intensity, phase and direction and reveal the presence of the conductors.

If the primary field is transient the secondary currents and their field will decay gradually when the primary field has ceased to exist. The decay is faster the higher the resistivity of the medium in which the currents flow. In this case we cannot talk of any unique phase relations since the signal contains an infinite number of frequencies. The discussion of transient methods will be deferred until Section 6.13.

A great advantage of the electromagnetic methods is that they can be successfully applied even when conductive ground connections, indispensable for the methods of the last chapter, cannot be made owing to highly resistive (or insulating) surface formations. This is frequently the case in arid tracts or in the polar and sub-polar regions where the ground may be frozen to a considerable depth.

On the other hand one of the troublesome effects in the electromagnetic methods is that the secondary currents in superficial layers of good conductivity, e.g. clays, graphitic shales, etc. may screen the deeper conductors partially or wholly from the primary field. The latter which are the real objects of exploration, will then produce weak or no distortions (anomalies) in the primary field and may therefore be indetectable.

6.2 Near and far fields

Consider, for example, a dipole source of electromagnetic waves such as a small circular coil (effective area A) carrying an alternating current $I_0 \exp(-i\omega t)$ of angular frequency $\omega (= 2\pi v)$ placed in a homogeneous, isotropic medium (resistivity ρ, dielectric constant ϵ, magnetic permeability μ). At any point in the medium there then exists an oscillating magnetizing force **H** whose components H_r, H_θ, along and perpendicular to the line joining the dipole and the point, are given by

$$H_r = \frac{AI_0}{2\pi r^3} (1 - ikr)e^{-i(\omega t - kr)} \cos \theta \qquad (6.1a)$$

$$H_\theta = \frac{AI_0}{4\pi r^3} (1 - ikr - k^2 r^2)e^{-i(\omega t - kr)} \sin \theta \qquad (6.1b)$$

where r is the distance from the dipole, θ the angle made by the joining line with the dipole axis and

$$k = (\omega^2 \epsilon \mu + i\omega\mu/\rho)^{1/2} = a + ib \text{ (say)} \qquad (6.2)$$

It can be seen from these equations that at small distances ($|kr| \ll 1$), the so-called *near-field or induction region*, the amplitude of **H** varies as $\exp(-br)/r^3$. In the *far-field or radiation region* ($|kr| \gg 1$) it varies as $\exp(-br)/r$. The field structure is more complicated in the *intermediate region*. There are, of course, no sharp boundaries between these various regions but a practical limit for the extent of the near-field region may be defined by $|k|r \leqslant 0.1$.

For practically all rocks, with a few notable exceptions such as water-logged soils, wet clays etc, ϵ is of the order of 90 pF/m while for μ we may put μ_0 (cf. Table 1). It can then be shown from Equation (6.2) that, if $v\rho < \sim 10^8 \ \Omega$ m/s, the near-field region may, for practical purposes, be taken as $r \leqslant 35.6(\rho/v)^{1/2}$ m.

The limit $v\rho < \sim 10^8 \ \Omega$ m/s above implies that displacement currents in the ground may be neglected ($\rho\epsilon\omega \ll 1$, e.g. $\rho\epsilon\omega \leqslant 0.05$) so that the field may be considered as 'quasi-static'. For the frequencies commonly employed in geophysical work (maximum about 20 kHz) the condition $\rho\epsilon\omega \ll 1$ is satisfied with a good margin in many, if not most, areas.

Electromagnetic fields are propagated with the velocity of light but in restricting ourselves to quasi-static fields we are assuming that a state corresponding to instantaneous propagation holds at every point of the field.

6.3 Phase and polarization

6.3.1 Phase relations

It will be convenient at the outset to recall briefly a few elementary ideas. It is well known that if a primary magnetic field $P = H_0 \sin \omega t$ of frequency $\omega/2\pi$ acts on an electric circuit, say a coil, the secondary induced e.m.f. lags $\pi/2$ behind the primary field. If R and L denote the resistance and the self-inductance of the coil, the current in the coil and the secondary magnetic field produced by it lag $\pi/2 + \phi$ behind the primary field, where $\phi = \tan^{-1}(\omega L/R)$, the lag $\pi/2$ being caused by the fundamental law of induction and ϕ by the properties of the secondary circuit.

It follows that a very good conductor produces a secondary field almost opposite in phase to the primary field ($R \to 0$, $\phi \to \pi/2$) while a bad conductor produces a field that lags 90° behind the primary one ($R \to \infty$, $\phi \to 0$). The same phase effects are obtained on increasing and decreasing the frequency of the primary field (or the inductance in the secondary circuit).

The relation between the primary (P), the secondary (S) and the resultant (R) fields is shown in Fig. 56 which is a conventional vector diagram of electric circuit theory. The component of S in phase with P (the real component) is $-S \sin \phi$; the component lagging 90° behind P (out-of-phase or imaginary component) is $-S \cos \phi$.

The real and imaginary components of the induced signal can be conveniently expressed in units such as millivolts, nWb/m² per ampere primary current, etc. but it is also customary to express them as fractions or percentages of the primary field amplitude.

The real and imaginary components of the secondary response of a single-turn loop, to a homogeneous sinusoidal field, are

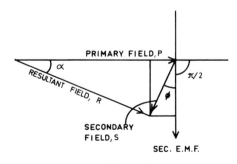

Fig. 56. Phase relations.

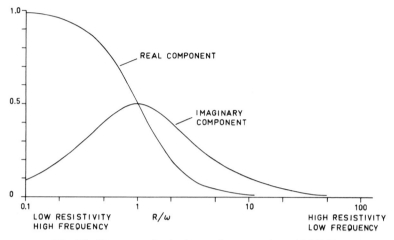

Fig. 57. Response of a single-turn loop to a sinusoidal field.

plotted in Fig. 57 against R/ω. This case reproduces qualitatively every essential detail of the induction response of complicated conductors to sinusoidal fields, homogeneous or otherwise.

Simple rules for identifying conductors may be formulated from Fig. 57. Thus, a good conductor produces a large real but a small imaginary component while a bad conductor produces a relatively large imaginary but a small real component. If the conductor has a 'medium' resistivity both the components are moderately large. Quantitatively, the ratio of magnitudes 'Real/ Imaginary' is often used, this being greater than 1 for good and less than 1 for bad conductors.

6.3.2 Elliptic polarization

Suppose that the plane of the coil in the above example is vertical and that the (homogeneous) primary field is horizontal. The resultant field at any point above the coil may be resolved into a horizontal (X) and a vertical (Y) component. It can be seen that

$$X = H_0 \sin \omega t + s \sin(\omega t - \pi/2 - \phi) = A \sin \omega t + B \cos \omega t$$

$$Y = s' \sin(\omega t - \pi/2 - \phi) = A' \sin \omega t + B' \cos \omega t \qquad (6.3)$$

where s, s' are the amplitudes of the horizontal and the vertical components of the secondary field.

Eliminating ωt from (5.1) we get

$$(A'^2 + B'^2)X^2 + (A^2 + B^2)Y^2 - 2(AA' + BB')XY$$
$$- (A'B - AB')^2 = 0 \qquad (6.4)$$

which is the general equation of an ellipse. This means that the resultant field at any point is elliptically polarized, the vector $\sqrt{(X^2 + Y^2)}$ describing an ellipse $\omega/2\pi$ times per second. In fact, the resultant field is always elliptically polarized irrespective of the nature of the primary field or the number or nature of the secondary circuits. The ellipse will degenerate into a straight line if the conductor is very good ($\phi = \pi/2$).

Referring to Equations (6.1) it should also be noted that in the near-field region the two orthogonal components H_r and H_θ are in phase with each other, the phase of either being *ar* with respect to the dipole current, so that in this region the field is linearly polarized, as is also the case in the far-field region (except for $\theta = 0$). However, in the intermediate region the field is elliptically polarized and the parameters of the ellipse (orientation, major axis and minor axis) vary from point to point.

6.4 Classification of continuous wave methods

Many early electromagnetic methods were based upon determining the azimuth and the dip of the ellipse of polarization as well as its major and minor axes which, it can be shown, determine the amplitudes of the real and imaginary components of the resultant field. It is easy, in principle, to find these parameters by means of a search coil which can be oriented in any desired direction.

If the coil is connected via an amplifier to a null detector the plane of the ellipse will be indicated by the position of the coil in which a null is obtained in the detector. Setting the coil normal to the plane of polarization and measuring the maximum and minimum signals induced in it as it is turned through one complete revolution, one obtains the two axes of the ellipse. Although the method provides the most complete information about the electromagnetic field at a point it is cumbersome to use and the desired accuracy is not easy to attain. In a number of other methods (the so-called tilt angle methods) only the tilt of the major axis of the polarization ellipse from the horizontal or the vertical is determined. However, tilt angles do not provide complete information about the electromagnetic field.

Since in most practical work the primary field is either vertical or horizontal tilt angles can be determined relatively easily as follows. In the former case the polarization plane must necessarily be vertical and is found by turning a coil around a vertical axis. The coil is then held perpendicular to the plane and turned around its horizontal diameter to obtain the minimum signal. The coil axis now coincides with the minor axis of the ellipse and its inclination with the horizontal is equal to the tilt of the major axis from the vertical.

The situation is slightly more complicated if the primary field is horizontal because, although the polarization plane will contain the primary field, it will not necessarily be vertical or horizontal. The plane can be found by turning the coil around an axis coinciding with the primary field. The coil is then set perpendicular to this plane and turned around an axis at right angles to the plane, until a minimum signal is obtained. The coil diameter perpendicular to the axis of rotation coincides in this position with the major axis of the ellipse of polarization.

In many other methods the real and imaginary components of the resultant field, rather than the geometrical parameters of the field, are determined. These as well as the tilt angle methods fall into two main categories:

(1) Methods in which the source of the primary field is stationary and the receiver arrangement mobile;

(2) Methods in which the source as well as the receiver is mobile.

The first category includes the so-called Compensator and Turam methods; the second includes methods or to be more accurate, measuring outfits or procedures, which go under various names such as Slingram, electromagnetic gun (EMG), demigun, Horizontal Loop systems, broadside and shoot-back techniques etc.

6.5 The Compensator or Sundberg method

The primary layout in this method consists of a straight cable, some 400 to 4000 (or more) metres long, grounded at both ends, through which an alternating current of low frequency (<1000 Hz) is passed. A large horizontal loop, usually rectangular, may also be laid on the ground instead of the cable. The cable (or the long side of the primary loop) is generally placed approximately parallel to the geological strike in the area and the electromagnetic field is investigated at regular intervals along lines perpendicular to it. In the case of the loop, observations can be made inside as well as outside the loop.

For many, if not most purposes, it would be sufficient to measure the amplitude and phase of either the vertical or the horizontal component of the resultant field, but measurements of both may be called for in detailed work. A search coil consisting of several turns of copper wire on a suitable frame is held horizontally (to measure the vertical component) or vertically (to measure the horizontal component) and the voltage induced in it is compared with a reference voltage. The latter is obtained from an auxiliary ('feeding') coil stationed near the primary layout. A method has also been tried in which the reference voltage is instead transmitted by an UHF carrier-wave modulated at the frequency of the primary current.

The comparison of voltages is made on a compensator, essentially an a.c. potentiometer, in which the inclusion of a reactive element enables one to determine phase differences. A simple circuit of this type is shown in Fig. 58. The real component is balanced by the voltage across the resistor R and the imaginary one by that across the variometer I, the balance being indicated by a silence in the headphones. For a discussion of other types of compensators see [113]. The real and imaginary signals in the receiver can be expressed after calibration as nanotesla per ampere primary current. In earlier papers, and the figures reproduced below from these, the field has been expressed as microgauss per ampere. It should be noted that $1~\mu G = 0.1$ nT. The phase of the received field with respect to the primary field is given by $\tan^{-1}(\text{Imag}/\text{Real})$. However, circuits can be devised that give a phase reading directly.

The normal primary field of the layout must be subtracted from the observed real component. The imaginary component needs no correction. These considerations are valid provided the field is linearly polarized. However, it is evident from Section 6.3.2 that at appreciable distances from the layout the field may be elliptically polarized and a simple correction for the normal field may not be sufficient.

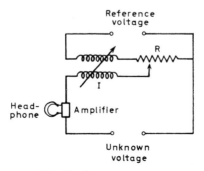

Fig. 58. A.c. compensator.

The general problem of delineating sub-surface conductors by electromagnetic methods is one of locating secondary current concentrations. Owing to the skin-effect the currents tend to be localized on the boundaries of a conductor. The most pertinent case, especially in such applications as ore prospecting, is that of a vertical sheet-like conductor representing an ore vein. Depending upon the position of such conductors with respect to the primary source they may be mapped according to the following considerations.

(1) If the vertical conducting sheet is cut transversely by the field of a long cable or that of a large loop the currents will evidently flow mainly along vertical surfaces, being strongest on the side nearest the primary source. The main features of the secondary magnetic field on the surface may be visualized by replacing the secondary current distribution by a vertical sheet having only transverse magnetization.

(2) If the sheet is *inside* a loop it will be subjected to a more or less uniform vertical field and the secondary currents will flow mainly in horizontal planes, being strongest on the upper face and decreasing downwards in depth. The secondary, surface magnetic field will resemble the field of a vertically magnetized sheet in its broad features.

Thus, the conductor will be located below the inflexion point of the secondary vertical field in case (1) and below the maximum in case (2). Fig. 59 shows the results of some laboratory model experiments (after Sundberg [114]) corresponding to the first case. It will be seen that the phase also changes in a characteristic manner along the line of measurements and, moreover, that the width of the conductor is almost exactly equal to the distance between the points at which the phase of the vertical and horizontal components is 180°.

Perhaps the most interesting application of the compensator method is in the determination of the depth and conductivity of a number of thin, horizontal conductors, one below the other but not necessarily contiguous. A sedimentary column with interspersed conducting beds, for instance, can be approximated by such a model.

The electromagnetic field at a point P above a thin horizontal conducting sheet when an alternating current flows through an infinitely long cable, parallel to the sheet and at a height h above it, can be calculated from a formula derived by Levi-Civita [115]. The field is a function of y, $h + z$ (Fig. 60) and the 'induction

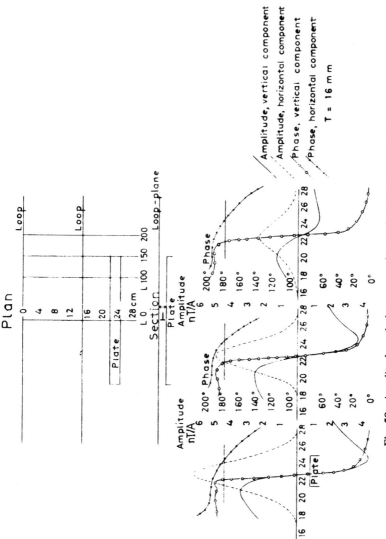

Fig. 59. Amplitude and phase anomalies in the Sundberg method.

149

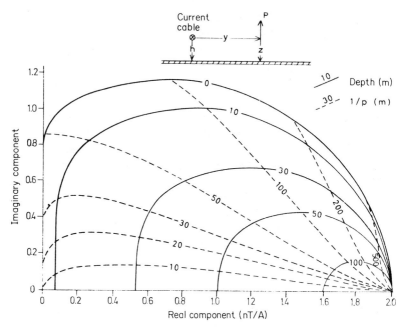

Fig. 60. Vector diagram for the secondary field due to a long current above a thin sheet conductor.

factor' of the conductor $p = \mu_0 \omega d/(2\rho)$ where d(m) is the thickness of the conductor. Evidently $1/p$ has the dimension 'metre'.

The real and imaginary parts of the vertical electromagnetic field above a horizontal conductor at a point distant 100 m from the primary current and at the same height as the current ($z = h$) are plotted on the vector diagram in Fig. 60. Here the solid curves show how the components vary for different conductors with different $1/p$-values but at the same depth and the dotted ones how they vary for a conductor with a given $1/p$-value but situated at different depths. A line drawn from the origin to a point on a solid curve represents the total vertical field vector (cf. R in Fig. 56).

It will be realized that a single in-phase and out-of-phase observation at a known distance suffices to determine the depth as well as the induction factor of the thin horizontal conductor.

Another important property of the vector diagram in Fig. 60 is that it can be used to construct the electromagnetic field above any number of thin, plane and parallel conductors, one below the other. The procedure may be illustrated by reference to the simple case of two conducting sheets 1 and 2, having induction factors p_1, p_2 placed at depths z_1 and $z_2 (>z_1)$. If the sheet 1 is immediately on top of sheet 2 the two form a 'combined' thin sheet at depth z_2 with an induction factor $p = p_1 + p_2$. The field of this sheet as well as the field of sheet 1 alone placed at depth z_2 can be read off Fig. 60. The vector difference between them gives the field of sheet 2 placed at depth z_2. This field can be added vectorially to the field of sheet 1 alone at depth z_1, also read off Fig. 60, and we obtain the field vector of the system of two thin sheets at depths z_1 and z_2. The theory of this construction and its generalization to any number of thin horizontal conductors has been discussed by Sundberg and Hedström [116].

In applying the above to deduce the unknown depths and induction factors of horizontal conducting beds from compensator observations the vector constructions are reversed. The total field vector is divided into its constituent parts and a solution obtained for the system of conductors. It should be realized that this is not a 'trial-and-error' procedure in the usual sense but one that gives a unique solution depending upon the number of observations available. This point has been very ably discussed by Sundberg and Hedström:

'Since every field vector measured means the determination of two coordinates in a vector diagram . . . , it will allow the determination of two unknowns, one depth and one induction factor. Every reading added, whether taken from another component of the field vector, at another distance from the primary cable or at another frequency, will therefore make it theoretically possible to determine two more unknowns. . . . In practical work it is necessary, however, in order to obtain a solution within reasonable time, to substitute a simple (but representative) system of conducting sheets, comprising not more than four or five unknown quantities . . .'.

The compensator method has been extensively used mainly for structural studies such as following up electrical 'key-beds' in oil field formations or following up the depth extension of flat-lying or gently dipping ore lenses [66, 113, 117].

6.6 The Turam method

The compensator method requires a direct connection between the primary layout and the observation point and therefore becomes cumbersome when large areas are to be covered. This operational disadvantage is overcome in the Turam method devised by Hedström [118].

The primary field is produced as before by a long cable or a large loop, and two search coils, 10–50 m apart, are carried along the line of measurements. For each position of the coils the ratio of the amplitudes of and the phase difference between the voltages induced in them are measured on a bridge type compensator, the former with an accuracy of about 0.01, the latter about $0.2°$.

The coils are usually held horizontally (so as to compare the vertical components of the resultant field) but may sometimes be kept vertical with their planes either parallel or perpendicular to the measuring profile. We shall restrict the discussion to the case of two horizontal coils.

The quantities measured in Turam work are V_1/V_2, $V_2/V_3, \ldots$, etc. and $\alpha_2 - \alpha_1$, $\alpha_3 - \alpha_2, \ldots$, etc. Where the V's are the amplitudes, and α's the phases of the vertical electromagnetic field at the stations 1, 2, 3.... To correct for the variation of the primary field (p) with the distance from the source the measured ratios are divided by the normal amplitude ratios p_1/p_2, $p_2/p_3, \ldots$, etc. This is particularly easy in the case of the long cable since the normal ratio of the vertical fields at any two points is simply the inverse ratio of the distances of the points from the cable, unless one or both of the coils are on a level different from the cable, when an easily calculable correction must be applied.

The normalized or reduced ratios $V_1 p_2/V_2 p_1$, $V_2 p_3/V_3 p_2$,..., etc. will all be equal to unity in the absence of sub-surface conductors.

The normal phase differences are, of course, zero provided the ground is non-conducting. This statement is not strictly true because the phase of an electromagnetic field varies by 2π over a distance of one wavelength. However, the air-wavelength of the low frequency primary field used in geoelectric work is of the order of several hundred kilometres. This is very long compared with the distance from the cable of a kilometre or so, up to which the measurements can usually be made so that the normal phase changes, although not strictly zero, are quite negligible.

On conductive ground, however, the phase may change appreciably within relatively short distances. Moreover, the field will be elliptically polarized, the inclination of the ellipse becoming more and more horizontal away from the cable. In accurate and detailed work these effects must be taken into account as otherwise uncertainties will be introduced in the assessment of the normal field [119, 120]. It is evident that deviations of the reduced ratio from unity and the phase difference from zero indicate anomalous sub-surface conditions.

Now, the phase difference measures essentially the horizontal gradient of the phase. As regards the reduced ratio it can be shown that, to a first approximation, it represents the function $1 - cs'/(1 + s)$ where s is the secondary field expressed as a fraction of the primary field at the point, s' its horizontal gradient and c the constant separation between the search coils. Thus, the *departures* of the reduced ratio from 1 constitute a measure of the horizontal gradient of the amplitude of the secondary field.

Typical vertical field indications obtained along three parallel Turam profiles across a sheet-like conductor are shown in Fig. 61. The conductor in question is a large pyrite ore body in an environment of porphyry rocks. It dips about 68° NE and the depth to its top surface is 14 m [121]. The frequency of the primary field was 540 c/s in this work and the source was at

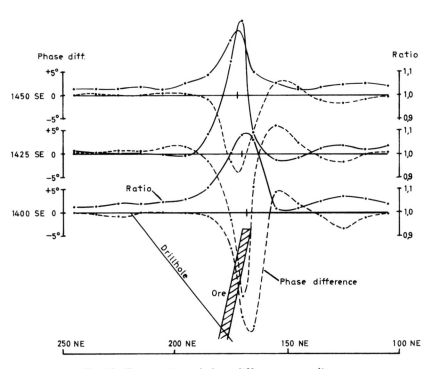

Fig. 61. Turam ratio and phase difference anomalies.

400 NE. The marked asymmetry in the curves is probably due to the dip of the conductor.

It is worth remembering that in case a sheet-like conductor has an attitude parallel to the primary field-lines at it, there will be no ratio and phase-difference anomalies above the conductor. Moreover, the signs of either indication will be reversed if the dip towards the source is shallower than this value, presuming throughout that the surrounding rocks are non-conducting.

It will be noticed that the ratio and phase difference curves show extremum values directly above the conductor and that they resemble the x-derivatives of the amplitude and phase curves for the vertical field in Fig. 59.

6.7 The moving source and receiver method (tandem outfits)

The layout of this method is shown in Fig. 62. A battery-operated portable oscillator $(1-2\text{ W})$ delivers an alternating voltage to a transmitter coil. The receiver is spaced at a fixed distance from the transmitter, usually between 25 and 100 m, and is connected to a compensator which is also fed by the reference voltage from the transmitter. The field acting on the receiver is measured in per cent of the primary field present at it when the system is on electrically neutral ground.

Appropriate corrections to the compensator readings must be applied to take into account the changes caused in the primary field at the receiver, due to a change (e.g. on account of topography) in the mutual separation, or orientation of the coils. These corrections, which are only needed for the real component, are easily calculated since the primary field being an ordinary dipole field is known everywhere. In rough terrain it is, however, preferable to sight the coils towards each other in such a way that their nominal orientation (e.g. coplanarity) is maintained, which

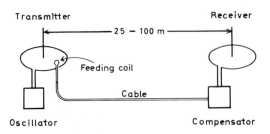

Fig. 62. Moving source and receiver system.

automatically eliminates the topographic correction to a large extent.

In principle, any mutual orientation of the transmitter and the receiver may be used but most surveys are carried out with both coils either horizontal or vertical. The transmitter–receiver layout is moved as a whole along their joining line and readings taken at suitable intervals. For horizontal coils and vertical coaxial coils the direction of movement is perpendicular to the geological strike (Fig. 63). For vertical coplanar coils it should be parallel to the strike. The response of a conductor to a moving source–receiver system is easily visualized in a qualitative manner by elementary arguments.

Consider, for instance, a long, *thin* vertical plate below the plane of the transmitter and the receiver, both of which are horizontal, the plate being perpendicular to the transmitter–receiver line (Fig. 63). Suppose that at a particular instant the oscillating transmitting dipole is directed downwards so that the primary field cuts the plate as shown. The secondary currents in the plate will be into the plane of the figure at the top edge of the

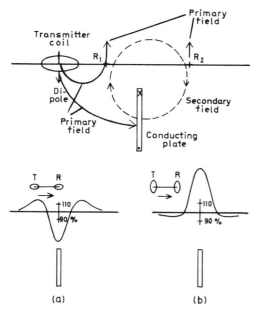

X - coordinate is the midpoint of the T-R line

Fig. 63. Origin of electromagnetic anomalies in the moving source–receiver system.

plate and out of it at the bottom. At any point R_1 on the same side of the plate as the transmitter the secondary field will be directed upwards, that is, in the same sense as the primary field at it. At a point such as R_2, however, the two fields will be in opposite directions and the resultant field will be less than the primary. Furthermore it is easy to see that if either the transmitter or the receiver is directly above the plate the field indicated by the compensator will be equal to the primary field.

Thus as the horizontal transmitter – receiver system with a fixed spacing is carried across the plate, the field picked up by the receiver will have the appearance as in Fig. 63a. The maximum damping in the field (negative anomaly) is obtained when the midpoint of the system is directly above the plate.

Similarly, it can be seen that if both coils are held vertically and are coaxial, the field will, under certain conditions, have the appearance as in Fig. 63b, there being a maximum *augmentation* of the primary field (a positive anomaly) when the plate is midway between the transmitter and the receiver. However, the response to this coil configuration varies in a rather complicated manner with the depth to the top edge of the plate and is generally difficult to predict correctly from simple arguments alone.

In Fig. 64 are shown the results of some small-scale laboratory experiments on a 1 mm thick zinc plate at different depths below the plane of the coils, these being horizontal and 100 mm apart. Fig. 65 is a simplified vector diagram on which are plotted the maximum real and imaginary component anomalies measured above several vertical conducting plates at two different depths (20 and 40 mm) and with different values for the factor $\rho/\nu d$ where ρ is the resistivity of the plate, d its thickness and ν the frequency. It will be noticed that, as in Fig. 60, a single field observation suffices to determine the depth and $\rho/\nu d$ for a *thin* plate and also that, for the same depth the field of all plates with the same $\rho/\nu d$ is practically identical. (The resistivities of the plates used are Cu 16.5, Al 29, Zn 60 and Pb 210 nΩ m.)

Experiments show that if the factor $\rho/\nu d$ is kept constant the thickness d must not exceed a certain value d_{max} if the field is to remain unchanged. This value depends upon ρ and ν. For a given ν, d_{max} is smaller the better the conductor and for a given ρ it is smaller the higher the ν. The result of this limitation on d is that the vector curve for each depth splits up as shown. This makes it possible, within certain limits, to estimate ρ and d separately by appropriately varying the frequency [122].

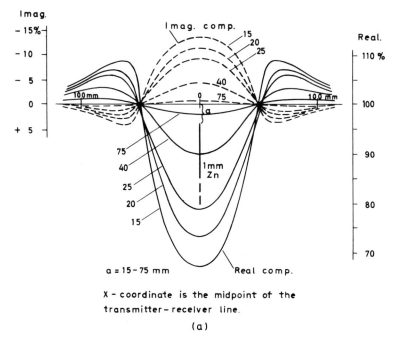

Fig. 64. Model electromagnetic experiments on a Zn conductor.

Vector diagrams as in Fig. 65 are of great value in the quantitative interpretation of field surveys. Sets of them for different depths, conductor dips, lengths etc. have been published by Nair *et al.* [123].

The response of *inclined* plates to the horizontal transmitter–receiver system is shown in Fig. 66a. In all these cases there is a minimum in the imaginary component (the *I* scale is negative upwards!) almost exactly above the top edge of the conductor. The minimum in the real component occurs at a point slightly towards the 'inside' of the plate. However, on a conductor poorer than that shown both minima are shifted towards the inside.

The *R* and *I* responses *along the line of measurements* are plotted in Fig. 66b as vector diagrams. The figure furnishes an interesting comparison of how the secondary vector 'swings' in the various cases as the coils are brought towards the conductor from infinity.

In Fig. 67 are reproduced the results of some field measurements on a thin sulphide conductor containing about 1 per cent

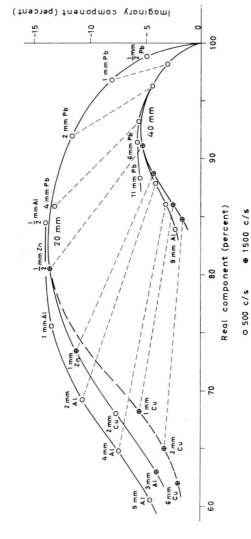

Fig. 65. Vector diagram for vertical conducting plates at two depths for the system in Fig. 63a. Coil distance = 100 mm. Note that the axis of the imaginary component is negative upwards.

158

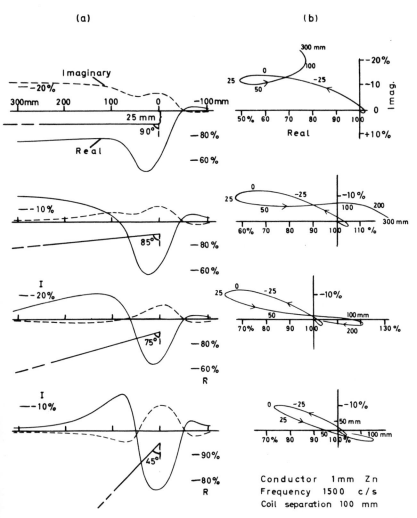

Fig. 66. Real and imaginary component anomalies with the system in
Fig. 63a across thin dipping sheets of great lateral extents.

copper as chalcopyrite, $CuFeS_2$. The conductor has a gentle dip of
15° with the horizontal. The measurements were made at two
different frequencies and two transmitter–receiver separations.
The smaller coil separation maps more detail in the anomalies but
the effect of near-surface conductivity variations is liable to be
accentuated. At the same time, however, the use of a lower
frequency emphasizes the relatively deeper conductivity variations. By a judicious application of such measurement techniques

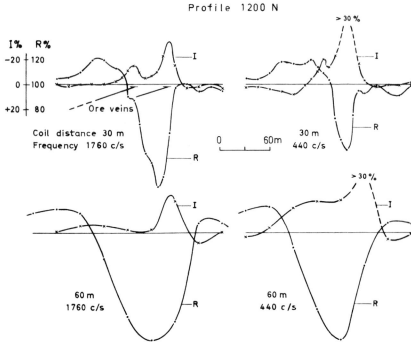

Fig. 67. Survey profiles with the moving source—receiver configuration in Fig. 63a.

it is possible to derive considerable detailed information about the conductors evinced by the electromagnetic anomalies obtained with the moving source—receiver system [122].

6.8 Broadside and shoot-back techniques

Both these techniques employ moving source and receiver but measure the tilt of the electromagnetic vector instead of the real and imaginary components, as in the techniques of the previous section. They have the advantage that no reference cable between the coils is required but, of course, the information obtained is less complete.

6.8.1 Broadside technique

In this case the transmitter (T) and receiver (R) are moved simultaneously along separate, parallel lines perpendicular to or at an angle to the geological strike. T is held with its axis horizontal and at right angles to the T—R line. R is turned around the T—R line as an axis until the signal in R is a minimum.

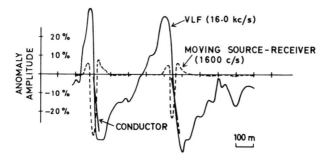

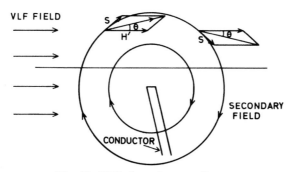

Fig. 68. VLF tilt-angle anomalies.

In the absence of conductors the field at the receiver, being the primary field only, will be horizontal. If conductors are present their secondary field will cause the resultant field vector to tilt out of the horizontal. The tilt, for certain geometries, will be in opposite directions on the two sides of a sheet conductor when the coil system traverses it, so that the tilt profile shows an inflection directly above the conductor (cf. Fig. 68). However, some quite complicated profiles can arise depending on the conductor depth.

6.8.2 Shoot-back technique

Rough topography may at times obscure the sight from one coil to the other, in which case the elimination of topographic effect by sighting (cf. Section 6.7) will not be feasible. In such cases, the shoot-back technique, devised by J. D. Crone, is useful in reconnaissance surveys. The procedure is as follows.

The coils are constructed so that either can be used as a transmitter or a receiver. The axis of coil 1 is directed towards 2, but downwards at an angle of 15° to the horizontal. Coil 1 transmits on the selected frequency and coil 2, acting as a receiver, is turned (for minimum signal) around a horizontal axis perpendicular to the line 1–2. The tilt α_1 of the field is noted. Thereafter coil 2 transmits but with the axis pointing 15° upwards. The tilt α_2 of the field is measured at coil 1, now acting as a receiver. In the absence of conductors α_1 and α_2 are very nearly equal regardless of the elevation difference between the coils so that any difference $\alpha_1 - \alpha_2$ is taken as a measure of the anomaly due to a sub-surface conductor.

6.9 Far-field methods

The reader may have gathered that the electromagnetic methods thus far described operate essentially in the near-field region. The effects of the intermediate region may at times manifest themselves, as has already been hinted at, but the far-field region is almost never involved in these methods.

The far-field dominates at distances beyond a few wavelengths from the source. In the methods dealt with hitherto the source is generally not powerful enough for measurements at large distances. However such measurements are possible in the field of strong sources and we now turn to these.

Powerful radio transmitters situated in different countries transmit unmodulated carrier waves, either continuously or with Morse code, for purposes of military communication, on frequencies in the band 15–25 kc/s. Some examples are: NAA, Cutler, USA, 17.8 kc/s, 1 MW; GBR, Rugby, England, 16.0 kc/s, 500 kW; ROR, Gorki, USSR, 17.0 kc/s, 315 kW; NWC, North West Cape, Australia, 15.5 kc/s, 1 MW. In radio technology these frequencies are known as *V*ery *L*ow *F*requencies (VLF), although it should be borne in mind that they are not low in the sense of the electromagnetic methods of applied geophysics, and the name is therefore improper in this connection.

6.9.1 H-mode VLF method

The antenna of a VLF transmitter constitutes a vertical electric dipole of moment iIl/ω (coulomb metre) where I is the (oscillating) current in it and l the antenna length. The magnetizing force (A/m) due to such a dipole in a homogeneous medium is horizontal and given by

$$H = \frac{Il}{4\pi} (1 - ikr) \frac{\exp(ikr)}{r^2} \sin \theta \qquad (6.5)$$

(cf. Section 6.2 where we considered a vertical *magnetic* dipole). Furthermore, H is perpendicular to the plane contained by the dipole and the radius vector *r*, which we shall call the *x*–*z* plane. H, therefore, has only the component H_y.

The field (H_0) at very large distances from the transmitter may, for all practical purposes, be considered as uniform within a small area and as having a direction perpendicular to the bearing of the transmitter from the observation point.

If α is the angle between H_0 and a long, sheet-like, steep conductor, the component $H' = H_0 \sin \alpha$ perpendicular to the conductor induces secondary currents in it, whose magnetic field (amplitude S_0) is not, in general, horizontal. Consequently, in the vicinity of the conductor the resultant field vector ($\mathbf{H_0} + \mathbf{S_0}$, assuming $\mathbf{H_0}$ and $\mathbf{S_0}$ to be in phase) is deflected above the horizontal on the side of the transmitter and below it on the opposite side (Fig. 68). However, $\mathbf{H_0}$ and $\mathbf{S_0}$ are not in phase, so that the field is generally elliptically polarized and in actual field work, therefore, not the vector $\mathbf{H_0} + \mathbf{S_0}$, but the tilt of the major axis of the polarization ellipse is determined.

If $\alpha = 0$, there is no induction in the conductor. Hence the choice of the VLF station must be made with due consideration to the geological strike.

Measurements are carried out as follows. A coil (or solenoid) tuned to the frequency of the selected VLF station is held with its axis horizontal and turned around a vertical axis so that a minimum signal is obtained. The vertical plane perpendicular to the axis in this position is the plane in which the major axis of the (magnetic) polarization ellipse lies. (The coil axis in this position will only indicate the bearing of the VLF station when far from any conductor.) Next, the coil is turned through 90°, the axis remaining horizontal, and finally it is tilted around the horizontal line through its centre and lying in its plane*, until a minimum signal is obtained. In this final position the major axis of the polarization ellipse lies in the plane of the coil.

It should be observed that the above procedure assumes that the plane of polarization is vertical. Unless this is actually the case, perfect null cannot be obtained with this procedure.

If ΔV is the amplitude of the vertical component of the secondary field due to a conductor, then it can be shown that†

---

*If a solenoid is used this line will obviously be the line perpendicular to the length of the solenoid.

†The relation is far from being intuitively obvious.

$2\Delta V \cos \phi / H$ = tan 2θ where H is the total horizontal field, θ is the tilt of the coil axis from the vertical (that is, of the major axis from the horizontal) and ϕ is the phase lag of the secondary field. Since usually $H \approx H_0$ and $\Delta V \ll H$ we can write $\theta \approx \Delta V \cos \phi / H_0$ so that $100 \times \theta$ (in radian) gives $\Delta V \cos \phi$ in per cent of the local primary VLF field strength H_0.

The upper part of Fig. 68 shows a profile across two parallel sulphide conductors, each several kilometres long, in N. Sweden. Also shown for comparison are the anomalies with the moving source−receiver method (1600 c/s). Further examples of H-mode VLF measurements will be found elsewhere [124−126].

6.9.2 Measurements of E and wave impedance

The electric field **E** of an oscillating electric dipole in an infinite, homogeneous medium has a somewhat more complex structure than the H-field represented by Equation (6.5). In general the field is elliptically polarized, the plane of the ellipse coinciding with the plane contained by the dipole direction and the radius vector. If the transmitter antenna is vertical (the z-direction) **E** at points on the horizontal plane through the dipole centre, the $x−y$ plane, has only a vertical component E_z.

In practice, we are not dealing with one homogeneous medium but two adjoining media (air and ground). In this case it can be shown that **H** still has only a component H_y at the ground surface but **E** has components E_z and E_x.

As far back as 1937 Norton showed [127] that on homogeneous, conductive ground, **E** is elliptically polarized and that, along the ground, at distances beyond about one wavelength from the antenna, the angle λ that the major axis of the ellipse makes with the vertical and the ratio minor axis/major axis remain virtually constant at all points. The exact expression for λ is complicated. Fortunately, since $\rho \epsilon \nu \ll 1$ (p. 142) in geophysical applications, it can be simplified and then gives

$$\tan \lambda = (\pi \rho \nu \epsilon_0)^{1/2}$$
$$= 0.527 \times 10^{-5} (\rho \nu)^{1/2}$$

where $\epsilon_0 = 8.854 \times 10^{-12}$ F/m.
It will be seen that even if ρ = 10 kΩ m and ν = 20 kHz, λ is barely 4.5 degrees, so that the major axis is practically vertical.

It can also be shown that beyond about one wavelength's distance on the ground surface,

$$| E_x |^2 = \frac{2\pi\mu_0 \nu^3 l^2 I^2 \rho}{c^2 x^2} \quad (\text{V m}^{-1}) \tag{6.6}$$

where $c \approx 3 \times 10^8$ m/s (velocity of light in vacuum) and x is the distance from the transmitter antenna.

Thus the measurement of $| E_x |$ will yield ρ for a homogeneous ground. On non-homogeneous ground we may define an *apparent resistivity* ρ_a by (6.6). Substituting numerical values,

$$\rho_a = 1.14 \times 10^6 \frac{x^2 | E_x |^2}{\nu^3 l^2 I^2} \quad (\Omega \text{ m}) \tag{6.7}$$

(x in km, E_x in μv/m, ν in kHz, l in m and I in A).

The mapping of ρ_a in an area in this way requires the knowledge of the dipole moment of the transmitter antenna and the distance x. However, the more complete solution of the problem shows that on homogeneous ground E_x and H_y differ in phase by 45° and that

$$| H_y |^2 = \frac{\nu^2 l^2 I^2}{c^2 x^2} \quad (\text{A m}^{-1}) \tag{6.8}$$

It should be noticed that $| H_y |$ is independent of ρ.

Combining Equations (6.6) and (6.8) it follows that

$$\rho = \frac{1}{2\pi\mu_0 \nu} \left| \frac{E_x}{H_y} \right|^2 \tag{6.9}$$

Quite generally, $E_x/H_y = 2\pi\mu\nu/k$, a complex quantity, namely the intrinsic wave impedance of the medium.

E_x can be measured by, say, a long, insulated wire antenna along the ground with a detector in its centre, and H_y by means of a coil antenna.

As with Equations (4.2) and (6.6), Equation (6.9) gives the true resistivity of a homogeneous medium, but for a non-homogeneous medium it defines an apparent resistivity.

Fig. 69 shows a ρ_a profile across a conductor. This was obtained from measurements of the wave impedance. The variation of the phase difference between E_x and H_y is also shown.

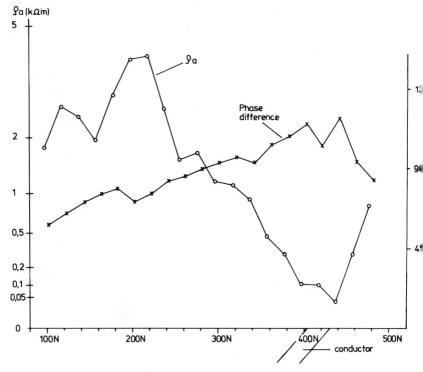

Fig. 69. ρ_a obtained from wave-impedance measurements across a conductor.
(The ordinate scale is quadratic.)

6.10 Theoretical approaches (continuous waves)

6.10.1 General

The problem of calculating theoretically the response of a conductor to an electromagnetic field is that of solving Maxwell's equations under appropriate boundary conditions. In the absence of any conductors the voltage induced in a receiver as a result of a current I in a primary source is $Z_0 I$ where Z_0 is the mutual impedance between the source and the receiver in vacuum. However, this voltage is also equal to KB_0 where B_0 is the flux density at the receiver. In the presence of a conductor the voltage will be $ZI = KB$, K being the same as before since it is a purely instrumental constant. Obviously, $B/B_0 = Z/Z_0$.

Thus, put in another way, the problem of finding the electromagnetic response is essentially that of calculating the mutual impedance of the source and the receiver in the presence of the conductor system of interest. In general the problem is extremely complicated and treatments were confined initially to simple conductor models such as a sphere [128–130], a semi-infinite earth [131], a cylinder [132], vertical and horizontal sheets [132, 134] and an earth whose conductivity is a function of the depth only [135, 136]. Very great progress has been made, however, in recent years in treating more complicated models such as a sphere enclosed by spherical shells [137] or a conductivity inhomogeneity in a layered half-space [138, 139] etc. The reader will also find two special issues of *Geophysics* [140, 141] illuminating from the theoretical point of view.

6.10.2 Electromagnetic sounding

Few of the conductor models mentioned above are very realistic geologically. One notable exception is the layered earth, and much interest has recently centred on the possibility of using the moving source–receiver dipole–dipole systems for electromagnetic sounding of horizontally layered structures.

Let us consider the system of Fig. 62 (coil separation r) on the plane surface of a semi-finite stratified earth. It can be shown [e.g. 142] that in the near-field region in this case we have

$$\frac{Z}{Z_0} = 1 - \int_0^\infty r^3 \lambda^2 R(\lambda, h, \rho, \nu) J_0(\lambda r)\, d\lambda \qquad (6.10)$$

where h, ρ are thickness and resistivity vectors $(h_1, h_2, \ldots, h_n;$ $\rho_1, \rho_2, \ldots, \rho_n)$. $R(\lambda)$ is a *complex* function corresponding to $K(\lambda)$ in Equation (4.9) and can be found for any stratification by means of recurrence relations. Similar equations exist for other loop configurations.

The real and imaginary parts of Z/Z_0 can be measured for different separations r and constant ν (*geometric sounding*) or for different frequencies ν but constant r (*parametric sounding*). The measurements thus obtained can then be fitted to a model of the stratified earth by calculating the right-hand side of (6.10). Kofoed *et al.* [142] have applied Ghosh's method (Section 4.6.3) towards this end and a more versatile approach along the same lines has recently been proposed by Johansen [143].

An examination of equations such as (6.10) shows that provided the ratio ρ/ν is not too small for a layer, the electromagnetic response depends on parameters like $h\sqrt{\rho}$ or $h/\sqrt{\rho}$ rather than on h and ρ separately. It will be recalled that for VES the corresponding parameters are $h\rho$ and h/ρ (Section 4.6.5). Thus, the interesting question arises whether the ambiguity due to equivalence can be overcome by combining VES measurements with VEMS (vertical electromagnetic sounding) ones. That this is possible to a very great extent was pointed out by Johansen in the paper cited above.

6.11 Model experiments

Although the formal solution of many of the theoretical problems may be available, the numerical computations are generally formidable and require recourse to a computing machine. Fortunately, however, the response of a 'natural' conductor can be exactly duplicated in the laboratory on a small, convenient scale. This is most easily seen from dimensional considerations.

Let the response of a conductor be measured in the dimensionless form S/P where S and P are the secondary and primary fields at a point. Since displacement currents are neglected the only relevant parameters producing this response are: $\rho(\Omega \text{ m})$, $\nu(\text{s}^{-1})$, the absolute permeability $\mu(\text{H/m} = \Omega \text{ s/m})$ and the linear scale of the experiment characterized by some length $d(\text{m})$. It is immediately verified that the quantity $\gamma = \rho/\mu\nu d^2$ formed from these four parameters is dimensionless. Therefore every system which has the same γ must produce the same dimensionless response S/P irrespective of the actual values of ρ, μ, ν and d*.

Suppose now that $d_{\text{full scale}}/d_{\text{laboratory}} = n$ and that the frequency of the laboratory experiment is the same as the 'full scale' experiment which it is supposed to represent. Then equating the γ's,

$$\left(\frac{\rho}{\mu}\right)_{\text{f.s.}} = n^2 \left(\frac{\rho}{\mu}\right)_{\text{lab.}} \tag{6.11}$$

Since, however, in both cases we are generally concerned with 'non-magnetic' conductors $\mu = \mu_0$, the permeability of free space, so that

$$\rho_{\text{f.s.}} = n^2 \rho_{\text{lab}}. \tag{6.12}$$

This means that if the linear dimensions in the laboratory experiment are n times smaller than those in the full-scale work, a laboratory conductor with resistivity ρ will correspond to a geometrically similar full-scale conductor of resistivity $n^2 \rho$.

---

*For a wider application of such considerations see "Dimensional methods and their applications" by C. M. Focken (Edward Arnold & Co.)

Thus suppose that 10^{-3} m (1 mm) in the laboratory is chosen to represent 1 m in full scale ($n = 10^3$). A transmitter–receiver separation of 100 m will then be represented by 100 mm, a conductor depth of 50 m by 50 mm and so on for all the other lengths involved. The response of a 1-mm thick zinc plate ($\rho = 6.0 \times 10^{-8}$ Ω m) will then be the same as that of 1-m-thick full-scale conductor, e.g. an ore vein, of resistivity $6.0 \times 10^{-8} \times 10^6 = 0.060$ Ω m.

The choice of model conductors is limited more or less to metal sheets and these do not show a very wide range of conductivities. However, since we have the scale factor, the frequency, and to some extent the magnetic permeability as additional variables at our disposal, a wide range of conductors in full scale can be electromagnetically simulated in the laboratory.

Model experiments have already been referred to in Sections 6.5 and 6.7. It should be easy now on the basis of the above discussion to 'translate' these to full scale.

6.12 Depth penetration

The question of how deep electromagnetic waves penetrate into the ground is of great importance in geophysics. If the ground were perfectly insulating the waves could penetrate to any distance. However, owing to the finite conductivity of most surface formations and of the underlying rocks the incident energy is absorbed and the amplitude of the waves decreases exponentially in traversing the conductors, due to absorption alone. In addition there will be a 'geometric decrease' as the wave spreads. This decrease depends upon the character of the source (cf. Equation 6.1). It is therefore convenient to discuss the topic of depth penetration with reference to plane waves, for which there is no such geometric decrease.

The amplitude of a plane wave is reduced to $1/e$ of its surface value within a distance $\delta = 1/b$ where b is defined in Equation (6.2). We shall call this distance the depth penetration. Two important extreme cases can now be distinguished.

If displacement currents are negligible ($\nu\rho < \sim 10^8$ Ω m/s) then, for 'non-magnetic' conductors ($\mu = \mu_0$)

$$\delta = 503.3 \left(\frac{\rho}{\nu}\right)^{1/2} \text{ metres} \tag{6.13}$$

which shows that for this case the depth penetration decreases with a decrease in resistivity and an increase in frequency.

If ρ = 2000 Ω m, a value typical of many glacial moraines, sandy clays, wet sandstones and chalk, the depth penetration will be about 700 m at ν = 1000 c/s and 225 m at ν = 10 000 c/s. However, owing to the presence of water containing dissolved salts many surface formations have resistivities as low as 100 Ω m and δ may be reduced to about 150 m even at ν = 1000 c/s. In sedimentary formations clays and shales with resistivities of the order of 1 Ω m are quite common and in such cases frequencies as low as 10 c/s will be required to obtain a depth penetration of 150 m.

The relative response of deep-seated conductors increases and that of shallow ones decreases as the frquency is lowered. This is illustrated by Table 10 constructed with the help of Fig. 60 for the case of two horizontal conductors, each 10 m thick but having different resistivities. The use of very low frequencies for deep exploration is however limited by the fact that the absolute magnitude of the signals decreases more or less in proportion to the frequency.

In the other extreme when displacement currents are appreciable ($\nu\rho > \sim 10^{10}$ Ω m/s) we have (again for non-magnetic media),

$$\delta = 1784\, \epsilon^{1/2} \rho \text{ metres} \tag{6.14a}$$

Taking ϵ = 90 pF/m as representative of most earth materials,

$$\delta = 16.9 \times 10^{-3} \rho \text{ metres} \tag{6.14b}$$

In this case δ is independent of frequency.

We see from Equation (6.14b) that if frequencies of, say, 10 MHz and higher are to penetrate more than a few tens of metres, the resistivity must be at least a few thousand ohm metre. Experiments have been reported in which radio and radar frequencies penetrated several hundred metres of rock. From the

Table 10   Relative frequency response of shallow and deep conductors

Conductor			Amplitude of secondary field (% primary field)	
No.	Resistivity Ω m	Depth m	1000 c/s	200 c/s
1	2.0	30	75.9	51.0
2	0.4	75	5.0	14.0

above it is clear that these rocks must have been highly resistive. In special circumstances (e.g. glacier ice) it may, of course, be possible to obtain a penetration of several kilometres even at radar frequencies [144].

It should be noted that if $\nu\rho$ lies between ~10^8 and ~10^{10} it is generally preferable to estimate δ from the exact expression obtained from Equation (6.2), namely

$$\frac{1}{\delta} = b = \left(\frac{\omega\mu}{2\rho}\right)^{1/2} [(\rho^2\epsilon^2\omega^2 + 1)^{1/2} - \rho\epsilon\omega]^{1/2} \qquad (6.15)$$

Apart from the depth penetration defined above there is another magnitude which is often referred to as the (practical) depth penetration. This is the maximum depth at which a conductor may lie and yet give a recognizable electromagnetic anomaly. This depth depends on the nature and magnitude of the stray anomalies ('noise') caused by near-surface conductivity variations, on the geometry of the deep conductor and on the instrumental noise. Experiments show that in the ideal case where the first noise type can be neglected, the maximum practical depth penetration of the various electromagnetic arrangements is between about 1 and 5 times the separation between the transmitting and receiving systems.

6.13 Transient-field methods (time-domain EM)

In recent years methods in which electromagnetic energy is supplied to the ground by transient pulses instead of by continuous waves have evoked considerable interest. A method known as 'Eltran' was tried in America in the early 1950's, in which electric transient pulses were applied to the ground through two current electrodes while the receiver consisted of two potential electrodes. A typical value for the spread of each pair of electrodes is 300 m, the distance between the pairs being relatively large. In such cases the transmitting system is virtually an electric dipole on the surface of a semi-infinite conductor. The pulse arriving at the receiver is distorted more or less depending upon the conductivity of the ground, the distance to the receiver, and the scattering it has suffered. Fig. 70 shows the distortion of a transmitted square wave pulse.

Of course, the energy in transient methods can also be supplied purely inductively by pulses in an insulated cable loop and the receiver can be a search coil as in the continuous wave methods.

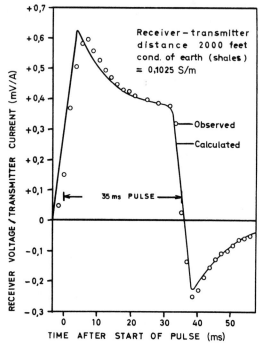

Fig. 70. Distortion of a square electric pulse due to transient-field phenomena [145].

The signal in the receiver decays gradually to zero as the secondary currents in sub-surface conductors dissipate on account of the electrical resistance, so that the smaller the resistivity of the ground, the slower is the decay of the received signal.

The transient and the continuous wave methods are related to each other through the Fourier transform (p. 257) according to which any waveform, transient or continuous, can be decomposed into a number of sine and cosine waves. Electromagnetic waves are delayed in passing through any conductor. In the case of sinusoidal waves this delay is measured as a phase shift while with pulses the actual delay-time is measured. It may be noted, for instance, that with a frequency of 1000 c/s a phase shift of 1° is equivalent to a delay of 1/360 x 1/1000, that is 2.78 μs. The transient signal received is a superposition of the secondary amplitudes of all the frequency components, each with its characteristic delay.

From Section 6.12 is evident that the high frequency components of a transient pulse are damped out at shallower depths than the low frequency ones. Consequently, the signals appearing at later times are governed by conductors lying at depths greater than the depths which govern the earlier signals.

On a non-homogeneous earth we can calculate an 'apparent conductivity', c_a(S/m), for each instant of measurement, defined as the conductivity that a homogeneous half-space must have to yield the secondary field actually observed at that instant with the given transmitter–receiver set-up. This concept of c_a as a function of time is due to Morrison *et al.* [146]. Fig. 71 shows c_a on a two-layer earth for a few different combinations of conductivities. We see that at early times the c_a curve represents the conductivity of the upper layer (σ_1) and at later times it approaches the conductivity (σ_2) of the lower layer (infinite sub-stratum) asymptotically, a situation entirely analogous to the variation of ρ_a with L in Section 4.4.

Transient methods posses certain advantages. It is clear from Section 6.12 that either very great primary field strengths at high frequencies or, if the field strengths are moderate, very low frequencies must be used in the continuous wave methods, if sub-surface conductors below highly conductive overburden are to be excited. In either case the instrumentation becomes very difficult, cumbersome and costly. The production of very high transient field strengths, on the other hand, is easier although the

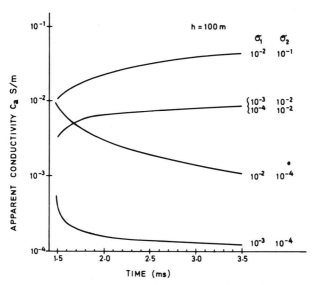

Fig. 71. Transient electromagnetic response of a two-layer earth.

difficulties involved should by no means be underrated. Experiments have been reported in which a power of 700 kW was dissipated in 100 ms through a square single-turn loop of side 300 m [147, page 342].

Several instances have been reported of sulphide ore bodies being detected by transient-field methods below highly conducting overburden [148, 149]. It has been suggested, though not explicitly, that such a detection is impossible by continuous wave methods. The problem deserves a more detailed experimental and theoretical study than has been made hitherto, in view of the theoretical equivalence between continuous wave and transient techniques referred to above.

Transient-field methods are still largely in the nascent stage but considerable theoretical work has been done in recent years on the responses of various types of conductors to transient pulses of different shapes [109, 150–155].

6.14 Natural-field methods

6.14.1 The telluric method

It has been known for a long time that electric currents of different kinds are flowing through the ground producing a potential difference between two points. Some of these currents are artificial, caused by electric railways, power lines, etc. while others may be natural, e.g. those due to the self-potentials of sulphide ores. Besides these local phenomena creating potential differences in limited regions there are currents caused by astronomic phenomena, such as the solar electron streams, the rotation of the earth, etc. which cover extremely large areas. These currents flow in vast sheets involving the entire surface of the earth and are therefore called telluric currents.

The telluric electric fields are of the order, say, of 10 mV/km and are constantly fluctuating in direction and magnitude at any point.

If X and Y are two orthogonal components of the telluric field at any point on the earth's surface they are related to the components X_0 and Y_0 simultaneously existing at a base point by the simple linear equations:

$$X = aX_0 + bY_0 \qquad (6.16)$$
$$Y = cX_0 + dY_0$$

The matrix $\| ab/cd \|$ is characteristic of the electric properties of the surface down to a depth of a few kilometres at the point under consideration.

On account of the linear form of Equations (6.16) the relationships between the time derivatives of the components are exactly the same as those between the components themselves. In the telluric method the variations ΔX, ΔY ΔX_0, ΔY_0 in successive intervals of, say, 10 s are determined from the simultaneous photographic records of the potential difference between two X and two Y electrodes at the field point and two X_0 and two Y_0 electrodes at the base.

It is easy to see that the successive normalized components $\Delta X_0/\Delta R_0$ and $\Delta Y_0/\Delta R_0$ where $\Delta R_0^2 = \Delta X_0^2 + \Delta Y_0^2$ define a circle while the vectors $\Delta X/\Delta R_0$ and $\Delta Y/\Delta R_0$ define an ellipse. The ratio of the area of the ellipse to that of the circle at the base is a convenient measure of the relative telluric disturbance at the field point.

Telluric currents obey the ordinary laws of electricity. Thus, for instance, if, E is the electric field at a point, then $E = \rho j$ (Ohm's law) where j is the associated current density. Consider a vertical contact between two media (Fig. 47) and imagine a telluric current density in the plane of the figure. If $\rho_2 < \rho_1$, then j normal to the contact will increase as we approach the contact from medium 1 and by Ohm's law E, that is $\Delta R/\Delta R_0$, will also increase. Conversely, $\Delta R/\Delta R_0$ will fall as we approach the contact from medium 2. E must therefore be discontinuous across the contact. Actually its variation is very similar to that of ρ_a in Fig. 47 or 48.

In general all geologic structures which tend to disturb the horizontal flow of the telluric current sheets, e.g. salt domes, folded strata, buried ridges, etc. will cause telluric anomalies. Detailed descriptions of the telluric method will be found elsewhere [156–159].

6.14.2 The magneto-telluric method

This is in a sense a further development of the telluric method. Briefly, it involves a comparison of the amplitudes and phases of the electric and magnetic fields associated with the flow of telluric currents.

The measurement of the electric field is relatively easy as indicated above, that of the magnetic field is considerably more difficult since we are concerned with frequencies around 1 c/s and down to 0.001 c/s or less. It is necessary to have coils with highly permeable cores and some 20–30 000 turns of wire. Such a coil can be about 2 m long and weigh some 30–40 kg. The voltages induced in the coil are detected by very high gain, low-noise amplifiers and the entire equipment for the measurement of the magnetic field can easily weigh some 70 kg or thereabouts.

Consider a telluric current sheet of frequency ν flowing in an electrically homogeneous earth of resistivity ρ. The depth penetration of such a sheet, that is, the depth at which the current density in it falls to $1/e$ of its value at the surface, is given by Equation (6.13).

It can be shown that the surface electric and magnetic fields are horizontal and orthogonal and that their amplitudes, E_x and H_y, are related by the same equation as Equation (6.9). Some workers prefer to consider the quantity measured by a magneto-telluric coil in air as the flux density B_y (Wb m^{-2} or T) rather than as the magnetizing force H_y (A m^{-1}). Since $B_y = \mu_0 H_y$ we get

$$\rho = \frac{\mu_0}{2\pi\nu} \left| \frac{E_x}{B_y} \right|^2 \tag{6.17}$$

Here E_x is in V m^{-1}. Expressing E_x in mV/km and B_y in nT we can write (6.17) as

$$\rho = \frac{0.2}{\nu} \left| \frac{E_x}{B_y} \right|^2 \tag{6.18}$$

where ρ is in Ω m. The phases of E_x and H_y (or B_y) differ by $\pi/4$, H_y lagging behind E_x.

If, then, we measure E_x and H_y at a definite frequency the first indication of the non-homogeneity of the earth will be that the phase difference θ will not be $\pi/4$. Secondly, ρ calculated from measurements at different frequencies will not be the same. However, we can always define an *apparent* resistivity ρ_a by Equation (6.18).

On determining ρ_a and θ as functions of frequency by actual measurements we obtain magneto-telluric soundings in a manner analogous to the electric soundings in Chapter 4 where, however, the current penetrates deeper because the electrode separation is increased.

Theoretical calculations of ρ_a and θ as functions of ν for a horizontally stratified earth have been made by Cagniard who has also given one of the classic accounts of this method [160].

Figs. 72a and b show ρ_a and θ for a two-layer earth. The major application of this method in the future is likely to be in elucidating very deep structures. In fact, soundings down to several tens of kilometres, and even a couple of hundred kilometres, have already been claimed. The utility of magneto-tellurics in shallow prospecting is probably limited, but one possible application may be in some areas with highly conductive overburden. It should be noted that, *with a given source–receiver separation*, the d.c. resistivity methods are handicapped by the short-circuiting effect of a high-conductivity near-surface layer

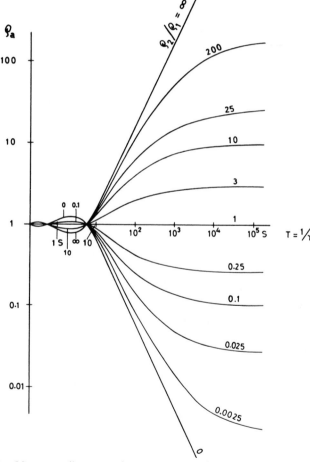

Fig. 72a. Magneto-telluric sounding: ρ_a against $T(=1/v)$ on a two-layer earth. This figure and Fig. 72b assume a first-layer thickness of 1 km and a resistivity of 1 Ωm.

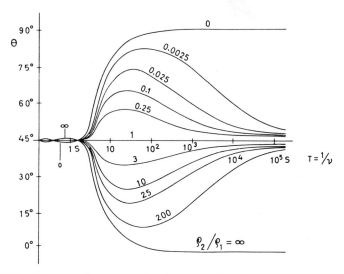

Fig. 72b. Magneto-telluric sounding: θ against $T(=1/\nu)$ on a two-layer earth.

whereas the magneto-telluric (and similar but artificial source) methods can 'see' through the layer to the required depth if one chooses an appropriately low frequency.

The reader should not, however, be unduly optimistic about the possibilities of the magneto-telluric method from the above, because usually the scatter of the observed ρ_a values prohibits an unambiguous interpretation of the ρ_a versus ν curves. Furthermore, the principle of equivalence (p. 119) also applies here.

One of the basic assumptions in the application of the telluric methods is that of uniformity of the fields ('plane wave approach'). There has been considerable discussion as to its validity but the fact that in areas with uniform sedimentary strata, magneto-telluric data are repeatable and independent of time has been interpreted to mean that the fields are uniform. It has also been pointed out [161] that ρ_a may depend on the direction in which E and H are measured and therefore a correct interpretation of sounding graphs in some cases requires a tensor analysis of the $\rho_a-\nu$ curves.

A good review of the theoretical and observational work in magneto-tellurics up to 1970 has been given by Keller [162] and the method has also been described in a coherent way in a recent monograph by Porstendorfer [163].

6.14.3 AFMAG

Natural magnetic fields of all frequencies from very low to very high ones are reaching all points on the earth. The frequencies that are exploited in the magneto-telluric methods are below about 1 c/s. The corresponding energy comes from complicated inter-actions between plasma emitted from the sun and the earth's magnetic field. Above 1 c/s the energy of the natural electro-magnetic fields seems largely to come from local and distant thunderstorms and man-made electrical disturbances. Depending upon local and seasonal variations the main part of their energy seems to be in the region of a few hundred c/s to a few kc/s.

The space between the ionosphere and the earth's surface acts as a waveguide for these fields with the result that their vertical component is normally very small. Their amplitudes and directions tend to be random or rather quasi-random. Normally a search coil at any point will show a marked horizontal plane of polarization for these waves and a diffuse azimuth in that plane. However, in the vicinity of highly conductive bodies the plane of polarization tilts out of the horizontal while the azimuth becomes more definite, and the presence of the conductor is thereby revealed. Usually the field strength shows an inflection point above the conductor and is flanked on the sides by a maximum and a minimum (cf. Fig. 68).

The name AFMAG is derived from the fact that the fields picked up by the search coil are *a*udio *f*requency *mag*netic fields. On account of the relatively high frequency the coil weight need be only a fraction of that required in the magneto-telluric method. The tilt measurements are made for two different frequencies, one high and one low, and the response ratio 'low/high' provides a measure of the conductivity of the conductor. It will be seen from the discussion in Section 6.3 that if $S \cos \phi$ or α is used as a measure of the response such a ratio is greater than 1 for good conductors and less than 1 for bad ones.

6.15 Influence of magnetic permeability

We have assumed so far that the magnetic permeability of conductors is equal to that of vacuum. This however is not justified for some natural conductors, e.g. magnetite ores, for which the relative permeability may be as much as 10.

From Section 6.11 it is evident that the effect of increasing the permeability n times is the same as that of increasing the conductivity $(1/\rho)$ n times. However, besides the field due to induced currents there is, in the case of magnetic conductors, a purely *magnetic* secondary field due to the oscillating induced magnetism. This latter field is in phase with the primary field and is, generally speaking, opposed to the real component of the induction part of the secondary field.

For example, for the single-turn loop of Section 6.3.1, the primary field is $P \cos \omega t$ and the real component of the secondary field due to the coil current is $-S \sin \phi \cos \omega t$. If we imagine a very small needle or bar (susceptibility κ) to be situated at the centre of the loop, the field due to its magnetic moment will be equal to a geometrical constant times $\kappa P \cos \omega t$, which is opposite in sign to the real component of the current-created field of the loop (the geometrical constant is evidently identical). The latter's amplitude increases with frequency while the magnetic moment magnitude is independent of frequency. There is therefore a critical frequency for a magnetic conductor above which its response is typically inductive (e.g. damping in the real component) and below which it is predominantly magnetic (e.g. with an appearance as for the field of a body statically magnetized in the direction of the primary field).

In Turam and moving source—receiver methods the effect of magnetic permeability generally manifests itself by 'distorting' the ratio or real component curves from their symmetric or near-symmetric form into curves of the antisymmetric type exhibiting an inflection point instead of an extremum above the current concentration. The converse is the case for the curves in the Sundberg method (Fig. 59) on magnetic conductors. These are 'distorted' to more or less symmetric types. In other instances anomalies which are 'normally' positive are rendered negative and vice versa (Fig. 98).

The influence of permeability must be carefully considered in field work when magnetic conductors are suspected in an area [164]. It is worth noting however that the magnetic permeability affects mainly the real component of the secondary electromagnetic field rather than the imaginary component.

6.16 Airborne measurements

The airborne electromagnetic methods are adaptations of the ground systems described so far. A large number of modifications exist but the principal types may be summarized as follows.

6.16.1 Continuous wave systems

6.16.1.1 *Helicopter and wing-tip*

In its typical version this system employs vertical, coaxial transmitter and receiver coils with the dipole axis parallel to the line of flight (cf. Fig. 63b). The coils are often compact ferrite-core coils with a diameter of approximately 10 cm and are mounted about 10–20 m apart at the ends of a very rigid boom installed underneath a helicopter. The boom is carefully designed to minimize the motion of the coils relative to each other so that spurious signals in the in-phase component are eliminated.

The system is flown at a height of about 45–50 m and the in-phase and out-of-phase field components at the receiver are continuously registered (usually in parts per million of the primary field at the receiver) on the automatic recorder. Typical anomalies from sub-surface conductors range from a few hundred to a thousand parts per million, and conductivities are estimated semi-quantitatively by the 'in-phase/out-of-phase' response ratio. The shapes of the anomalies can be visualized by the same type of arguments as for ground systems (p. 155 or p. 163).

The coils can also be mounted on the wing-tips of a small aircraft. In that case they are situated in the same vertical plane (instead of being coaxial as in the helicopter system) with their axes in the flight direction.

The practical depth penetration of the two systems is of the order of 20–30 m below ground level (about 75 m below flight level).

6.16.1.2 *Dual frequency phase shift method*

The fixed-wing aircraft has certain advantages (longer range, higher survey speed, greater pay-load, etc.) over the helicopter. However, the desired constancy of the coil distance (less than about 1 cm in about 15–20 m) is difficult to achieve with the coils on the wing-tips.

Now, the out-of-phase component of an electromagnetic field is a purely secondary phenomenon and is independent of the variations in the coil distance*. In the phase shift system only the out-of-phase component is recorded at two frequencies, one low

---

*This is strictly true of the phase-shift and only approximately (but to a very high degree) of the out-of-phase component.

(e.g. 400 c/s) and one high (e.g. 2300 c/s). The response ratio 'low/high' instead of the ratio 'in phase/out-of-phase' provides a measure of the conductivity of the anomalous body.

The transmitter is at the aeroplane while the receiving coils is towed behind in a 'bird' at the end of a cable, some 150 m long. The plane must be flown at a height of at least 120–150 m in order that the bird may not hit the tree tops and get lost. Over undulating terrain and in 'bumpy' weather it might be necessary to fly it still higher. Measurements of the in-phase component are out of the question due to the violent movements of the receiver bird.

On account of the comparatively high altitude at which such a system must be flown the signals of deeper conductors fall to very low levels and are often no greater than the signals due to lateral variations in surface conductivity. In this sense, then, the depth penetration of the method is poor unless the system is flown very low, which however is rarely possible for technical reasons.

The out-of-phase component of very good conductors varies little with frequency and such conductors cannot be detected by phase-shift measurements alone. This disadvantage may, however, be offset in areas with conducting host rock and/or overburden because good conductors may collect the phase-displaced currents induced in the rock or the overburden and give rise to a measurable electromagnetic field. (This is often the reason behind the strong imaginary component anomalies obtained over grounded rails, metal pipes etc., which, in themselves, represent almost infinitely good conductors to electromagnetic prospecting systems. Their intrinsic imaginary component responses are zero.)

The system has been discussed in more detail elsewhere [165, 166].

6.16.1.3 *Rotating field*

This method was devised in Sweden with a view to overcoming the disadvantages of the systems under Sections 6.16.1.1 and 6.16.1.2, namely the small depth penetration and, in the second also the inherent impossibility of in-phase measurements.

The transmitter of the system consists of two coils, one horizontal and the other vertical, fed by alternating currents of the same amplitude and frequency but with a phase difference of 90°. The primary electromagnetic field at any point in the surrounding space is therefore a rotating elliptically polarized field.

The receiver consists likewise of two mutually perpendicular coils and is placed at a distance, say a, from the transmitter

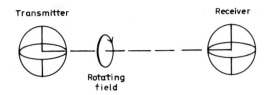

Fig. 73. Principle of the airborne rotating field system.

(Fig. 73). Along the line joining the transmitter and receiver in Fig. 73 the ellipse of polarization is a circle.

It is evident that if m is the dipole moment of either transmitting coil the primary field acting on the horizontal as well as the vertical receiving coil has an amplitude m/a^3. The voltages induced in the two receiver coils can be balanced against each other after shifting the phase of one of them, say that of the vertical coil voltage, by $90°$ so that the reading of a meter or a recorder is normally zero.

When a secondary field from a sub-surface conductor acts on the receiver there is an imbalance since the field will in general affect the two receiver coils unequally. The net voltage in the vertical coil is added (after shifting its phase by $90°$) to the voltage in the horizontal coil. The resultant has two components, one in phase with the primary field in the horizontal receiving coil and the other out of phase with it. These unbalanced components can be measured in terms of volts, amperes, tesla or any other convenient unit by suitably calibrating the deflection of the recorder connected to the receiver. It is, however, preferred to express them as percentages of the primary voltage induced in either receiver coil when a has a standard, specified value a_0.

It will be noticed that in contrast to the other airborne systems the rotating field system does not require a direct cable connection between the transmitter and the receiver. Also, the balance of the primary fields is independent of the variations in a since these affect the fields on both receiver coils in exactly the same way. It is therefore possible to place the transmitter in one aeroplane while the receiver is towed on a short cable (about 15—30 m) from another plane, flying in tandem at a distance of about 150—300 m. Such a system can be flown fairly low (about 60—80 m) and is not so dependent on weather conditions as the dual frequency system with its long towing cable.

Theoretically, the low flight height and the large transmitter—receiver distance should combine to give the system a depth range of more than 100 m below ground [167, 168]. However, the response of superficial horizontal conductors (e.g. soil layers) to the rotating field system is often very complicated, masking the anomalies of deeper conductors and making the interpretation difficult. According to tests in areas of high surface conductivity (East Africa) the practical penetration of the rotating field system in these areas appears to be only slightly better (40–50 m) than that of the wing-tip systems [169] although in areas of low surface conductivity (e.g. some pre-Cambrian shield areas) the predicted figure may perhaps be approached.

6.16.1.4 *Turair*

This system is based on the principle of the Turam ground method. It is a semi-airborne system in that the transmitter (large loop or long cable) is laid on the ground and traverses are made across it with the two Turam receiver coils towed from an aircraft. The corrections for the normal ratios are made by graphically removing the 'regional' gradient that appears on the profile record. They could, of course, be made by computation if the aircraft position with respect to the transmitter is known, but this would probably add greatly to the cost of the method. The interpretation of the anomalies obtained is exactly as in ground work, a remark that applies to all airborne work.

6.16.2 AFMAG

This system (Section 6.14.3) has also been adapted to airborne work [170]. A receiver consisting of two mutually perpendicular search coils, each making an angle of 45° with the horizontal, is towed behind an aircraft with the axes of the coils and the line of flight in the same vertical plane.

The flight direction is chosen perpendicular to the geologic strike in the region. Since the natural magnetic fields tend to be polarized perpendicular to the geologic strike the mean polarization vector then lies in the vertical plane through the flight line and the coil axes. If the vector has a tilt α from the horizontal, the voltages induced in the coils are proportional to $\cos(45° - \alpha)$ and $\cos(45° + \alpha)$. The difference of the two voltages is recorded in such a way that the deflection of a pen is approximately proportional to α. As in the ground AFMAG the tilt is recorded at two frequencies (150 and 510 c/s) so that relative estimates of the conductivities of anomalous conductors can be made.

6.16.3 Radiophase (VLF)

This is a system for combined *H*- and *E*-mode measurements from the air on the VLF band (Section 6.9). A block diagram of the system is shown in Fig. 74.

There are two electrical antennae, one vertical ('whip') and the other horizontal ('trail antenna'). Similarly there are two mutually orthogonal coils for picking up the magnetic field.

Since the amplitude of the vertical component (E_z) of the electric vector **E** is 100–200 times larger than that of the horizontal component (E_x), small departures from orthogonality between the two antennae cause large variations in the amplitude of the total signal picked up by the trail antenna. Hence a direct measurement of the ratio between the whip and the trail antenna voltages is out of the question. What is measured instead is the amplitude of the component of the trail antenna voltage that is 90° out-of-phase with the whip antenna voltage, in terms of the latter's amplitude. It is easily seen that the measured quantity is $(E_x \cos \gamma / E_z) \sin \phi$ where ϕ is the phase difference ($\approx 90°$) between E_x and E_z, and γ is the instantaneous angle made by the trail antenna with the horizontal. It is not possible to relate this quantity to the tilt of the polarization ellipse without a knowledge of γ, and since this is impossible in airborne work, we cannot calculate an apparent resistivity. The method is therefore only of semi-quantitative use for mapping conductivity variations.

The *H*-mode measurements are carried out in a similar manner.

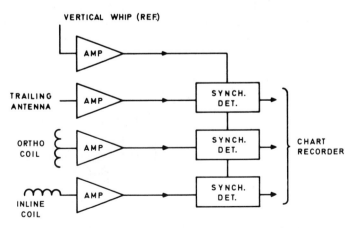

Fig. 74. Block diagram for the airborne Radiophase VLF system.

The components of the signals in the two coils that are 90° out of phase with the whip antenna voltage are measured and r.m.s. summed to give a measure of the amplitude of the secondary magnetic field. The measure is not theoretically exact.

As in ground work, the scope of the airborne VLF method is primarily limited, on account of the high frequency involved, to the mapping of superficial conductivity variations e.g. of soils, moraine cover, permafrost areas etc.

6.16.4 Airborne transient field (INPUT)

Transient pulses have been employed for aerial electromagnetic work. In the *in*duced *pu*lse *t*ransient (INPUT) system a wire is strung from the aircraft nose, around each wing-tip and beneath the tail, thus forming a large loop transmitter. Pulses of some 2 kW are sent through this horizontal loop at a repetition rate of several hundred times a second. A small 'bird' towed on a cable about 150 m long houses the coil which receives the secondary decaying signal from the ground. The signal is sampled (between the transmission pulses) at several pre-selected decay times (channels) from about 100 μs after current cut-off to about 2000 μs. The results can be expressed as ratios of the signals on different channels to, say, the signal in channel 1. A rapid decay (low late-channel signal) indicates a poor conductor in the ground, a slow decay a good conductor.

An advantage of the INPUT system is that since measurements are made during the time the primary field is cut off, the transmitter–receiver orientation geometry is of no consequence. Hence the sensitivity of the (broadband) receiver can be increased very considerably so that extremely weak signals can be detected. However, at the same time, we have no absolute reference for the strength of the recorded signals. In this the INPUT system resembles purely phase-measuring systems above. It is, in principle, a multi-frequency phase-measuring system, the later sampling times corresponding to successively lower frequencies. The lowest frequency that is effectively exploited cannot, of course, be less than the repetition rate of the pulse and the highest cannot be greater than the reciprocal of the time (after power cut-off) at which channel 1 is sampled.

The details of an INPUT instrumentation have been described by Gupta Sarma *et al.* [171] and some theoretical aspects have been treated by Verma [149, 172].

6.17 Note on the design of electromagnetic coils

The design of electromagnetic coils is a matter of some complexity. Here only the barest principles can be touched upon. We shall consider only flat, air-cored, circular coils (diameter D) for the moving source–receiver system of Fig. 62 with a separation (r) of 60 m, and derive the requirements for power (P), diameter (d) of winding wire, number of turns (N) in a coil, etc.

Suppose the r.m.s. noise field strength in an area (due to atmospheric disturbance, man-made disturbance etc.) is 1.6×10^{-8} A/m within a bandwidth of 0.1 Hz around a measuring frequency $f = 3600$ Hz. If a signal-to-noise ratio of 250:1, say, is desired at the receiver the r.m.s. normal field (H_0) at it should then be 4×10^{-6} A/m.

Now for the coil configuration in question

$$H_0 = \frac{NIA}{4\pi}\frac{1}{r^3} \quad \text{(A m}^{-1}\text{)} \tag{6.19}$$

where I is the r.m.s. current in the transmitter (area A). For a coil of reasonable size, say $D = 0.60$ m, we get then

$$NI = 39 \text{ ampere turns} \tag{6.20}$$

Suppose the transmitter is wound of Cu wire (resistivity ρ, density δ). Then if R is the resistance of the coil

$$P = RI^2 = \frac{\rho N \pi D I^2}{(\pi d^2 / 4)} \quad \text{(W)} \tag{6.21}$$

$$M = \delta N D \pi \frac{\pi d^2}{4} \quad \text{(kg)} \tag{6.22}$$

Hence

$$P = \rho\delta(NI)^2 D^2 \pi^2 / M \tag{6.23a}$$

Putting $\rho = 1.7 \times 10^{-8}$ Ωm and $\delta = 8930$ kg/m³ we get

$$P = \frac{0.82}{M} \quad \text{(W)} \tag{6.23b}$$

Usually $R \ll 2\pi f L$ where L is inductance of the coil so that if V is the r.m.s. supply voltage

$$I = \frac{\sqrt{2}V}{2\pi f L}$$

which can be written as

$$N = \frac{\sqrt{2V}}{2\pi f N I} \cdot \frac{N^2}{L} \tag{6.24}$$

The value of N^2/L can be found from inductance tables (e.g. Jahnke and Emde, Tables of Functions, Dover, New York, pp. 86–89). For a flat coil of length, say, 60 mm and $D = 0.60$ m we find from the tables that $N^2/L = 8.33 \times 10^5$.

Suppose $V = 30$ V; then for the example under consideration $N = 40$ turns. If we choose $M = 1$ kg it is then easily shown that $d = 1.4$ mm. For a coil of this weight (and the remaining specifications as above) we then get $P = 0.82$ W from (6.23b). Also, it follows that $L = 1.9$ mH and $Q = 2\pi f L/R = 52$.

It is preferable to construct *the receiver coil* with the same specifications as the transmitter coil. The minimum induced voltage in the receiver for the above example will be 1.3 μV ($-i\omega\mu_0 N H_0 A$) before tuning the coil, and 52 x 1.3 = 66 μV after tuning.

Here we have only considered the principles. For a discussion of the optimum design of e.m. sensing coils the reader may be referred to an article by Becker [173].

7 Seismic methods

7.1 Introduction

The seismic methods of geophysical exploration utilize the fact that elastic waves travel with different velocities in different rocks. The principle is to initiate such waves at a point and determine at a number of other points the time of arrival of the energy that is refracted or reflected by the discontinuities between different rock formations. This then enables the position of the discontinuities to be deduced.

The importance of the seismic methods lies above all in the fact that their data, if properly handled, yield an almost unique and unambiguous interpretation.

The standard method of producing seismic waves is to explode a dynamite charge in a hole. Attempts have been made to obtain the seismic energy by other means, e.g. weight-dropping, electrodynamic shaking, etc. but the energies thus produced are not generally sufficient and the methods have found only a limited application. In recent years, however, there has been a general tendency towards replacing explosive sources by such means.

We can classify the present-day seismic sources as follows: (1) Solid chemical explosives (Dynamite, Aquaseis, Flexotir, Primacord, Maxipulse are some of the trade names), (2) Compressed air sources (PAR, Seismojet, Terrapak etc.), (3) Electrical energy sources depending for their action on the sudden movement of a piston or a plate by transducer devices (Boomer, Pinger, Sono Probe, Sparkarray etc.), (4) Gas exploders (Acquapulse, Dinoseis, Deltapulse etc.), (5) Mechanical impulse sources (Hammer, Thumper), (6) Implosive sources (e.g. Hydroseis) and (7) Vibratory sources (Vibroseis).

Of these, Vibroseis falls in a class by itself on account of its continuous input signal. It is treated separately later on in Section

7.12. The others produce 'conventional' sudden pulses of short duration and the choice depends only on the operational requirements of the survey.

7.2 Elastic constants and waves

7.2.1 Hooke's law

The basis of the seismic methods is the theory of elasticity. The elastic properties of substances are characterized by elastic moduli or constants which specify the relation between the *stress* and the *strain*. A stress is measured as force per unit area. It is compressive (or tensile) if it acts perpendicular to the area and shearing if it acts parallel to it. A system of compressive stresses changes the volume but not the shape of a body, one of shearing stresses changes the shape but not the volume.

The strains in a body are deformations which produce restoring forces opposed to the stresses. Tensile and compressive stresses give rise to longitudinal and volume strains which are measured as the change in length per unit length or change in volume per unit volume. Shearing strains are measured as angles of deformation. It is usually assumed that the strains are small and reversible, that is, a body resumes its original shape and size when the stresses are relieved.

Hooke's law states that the stress is proportional to the strain, the constant of proportionality being known as the elastic modulus or constant. The law is not strictly true and more general stress—strain relationships have also been introduced in applied seismology, notably by Ricker [174, 175]; yet, Hooke's law carries us a long way in the theory of elasticity.

The two moduli of immediate interest for the study of the

Table 11   Elastic constants

Substance	Bulk modulus ($N/m^2 \times 10^{-10}$)	Shear modulus ($N/m^2 \times 10^{-10}$)
Marbles and limestones	3.7–5.7	2.1–3.0
Granites	2.7–3.3	1.5–2.4
Sudbury diabase	7.3	3.7
Ohio sandstone	1.25	0.61
Iron (wrought)	16.0	7.7
Iron (cast)	9.5	5.0
Glass (crown)	5.0	2.5
Quartz (fibre)	1.5	3.0

elastic waves in the earth are the bulk modulus (k) and the shear modulus (n). Their values for some rocks and for a few common substances will be found in Table 11.

7.2.2 Elastic waves

If the stress applied to an elastic medium is released suddenly the condition of strain propagates within the medium as an elastic wave. There are several kinds of elastic waves:

(1) In the longitudinal, compressional or *P* waves the motion of the medium is in the same direction as the direction of wave propagation. These are, in other words, ordinary sound waves. Their velocity is given by

$$V_1 = \sqrt{\left(\frac{k + (4/3)n}{\rho}\right)} \tag{7.1}$$

where ρ is the density of the medium.

(2) In the transverse, shear or *S* waves the particles of the medium move at right angles to the direction of wave propagation and the velocity is given by

$$V_t = \sqrt{\left(\frac{n}{\rho}\right)} \tag{7.2}$$

It is evident that $V_t < V_1$.

Shear waves can be polarized in which case the particles oscillate along a definite line perpendicular to the direction of wave propagation.

(3) If a medium has a free surface there are also surface waves in addition to the above two which are 'body waves'. In the Rayleigh waves the particles describe ellipses in the vertical plane that contains the direction of propagation. At the surface the motion of the particles is retrograde with respect to that of the waves. The velocity of Rayleigh waves is about $0.9\ V_t$.

(4) Another type of surface waves are the Love waves. These are observed when the velocity in the top layer of a medium is less than that in the sub-stratum. The particles oscillate transversely to the direction of the wave and in a plane parallel to the surface. The Love waves are thus essentially shear waves. Their velocity for short wavelengths is equal to V_t in the upper layer and for long wavelengths V_t in the sub-stratum.

The spectrum of the body waves in the earth extends from

about 15 c/s to about 100 c/s; the surface waves have frequencies lower than about 15 c/s.

In applied seismology only the *P* waves are of importance. In principle, however, *S* waves could also be used but the difficulty is to get *S* waves of sufficient energy. Explosions, which are the common means of generating powerful elastic waves, produce predominantly, if not exclusively, *P* waves. These are converted in part to *S* waves on oblique reflection and although some work has been reported with such *S* waves they have not been put to any great use in applied geophysics.

Surface waves are incapable of giving information about structures at depth and little interest is attached to them in exploration geophysics.

7.2.3 Velocities

Typical values for the velocity of *P* and *S* waves in some rocks are given in Table 12. The velocities are generally found to be greater in igneous and crystalline rocks than in sedimentary ones. In the sedimentary rocks they tend to increase with depth of burial and geologic age. Many empirical attempts have been made to represent this increase. For shales and sands Faust [176] finds that

$$V = 46.5(ZT)^{1/6} \text{ m/s} \tag{7.3}$$

where Z is the depth in metres and T the age in years.

Seismic velocities can be measured in the field as well as in the laboratory on samples of rocks. Several methods for laboratory determinations using magnetostrictive pulses, ultrasonic pulses, resonances etc. have been developed [177–180]. Well-velocity surveys are also a common method of obtaining information on velocities. A seismometer (Section 7.3.1) is lowered into a borehole, a shot is fired close to the surface and the travel time to

Table 12   Elastic velocities (m/s)

Material	Compressional	Shear
Air	330	—
Sand	300–800	100–500
Water	1450	—
Glacial moraine	1500–2700	900–1300
Limestones and dolomites	3500–6500	1800–3800
Rock salt	4000–5500	2000–3200
Granites and other deep- seated rocks	4600–7000	2500–4000

the seismometer is noted. The seismometer is then lifted a short distance and another travel time is measured. From the difference in these times the average velocity in the material between the two positions of the seismometer can be calculated.

Seismic velocities often show anisotropy in stratified media; the velocity parallel to the strata is generally greater than that normal to them by an amount of the order of 10–15 per cent.

7.3 Instruments and field procedure

7.3.1 Geophones and recording system

In the early days seismic waves were detected by a mechanical seismometer which consisted, in principle, of a heavy mass suspended as a pendulum. The mass remained stationary on account of its inertia while the suspension frame moved with the earth. Such mechanical devices have now been completely replaced in applied geophysical exploration, by small, lightweight electric detectors or *geophones*.

The simplest and the most common type is the electromagnetic geophone. A coil attached to a frame is placed between the poles of a magnet which, in turn, is suspended from the frame by leaf-springs. The magnet acts as an inertial element while the coil moves with the earth. The relative motion of the coil and the magnet produces an e.m.f. proportional to the *velocity* of the earth's motion. In underwater detection the geophones used produce a signal proportional to the excess *pressure* on a diaphragm. There is a very large variety of geophones on the market for land as well as underwater use, some of them weighing no more than 50 g.

The natural frequencies of electromagnetic geophones in reflection seismic work are usually 20 c/s or higher but may be as low as 5 or 2 c/s in refraction work. These frequencies are damped electromagnetically or by using oil, the damping being critical or slightly less than critical.

The voltage from the geophone is fed into an amplifier, usually over a transformer for reasons of impedance matching, and is then passed on to a galvanometer. Most amplifiers are provided with automatic gain control and a number of high-pass and low-pass filters which allow a selection of the frequencies to be recorded.

The galvanometer deflections are registered by means of a mirror and light-source system on a continuously running photographic film. A single film may carry signal traces from as many as 48 geophones. The record is called a *seismogram*.

Since each geophone needs an amplifier and a galvanometer of its own, compactness is of utmost importance in the design of seismic recording equipment. Indeed, the galvanometers in seismic cameras are only a few millimetres in diameter and a few centimetres in length, while the use of transistors has similarly led to amplifiers of small dimensions.

Vertical time lines, generally running across the entire width of the film, are superposed on the record. These are generated by a light beam which passes intermittently through a slot in a rotating disc whose speed is accurately controlled by a suitable device such as a tuning-fork driven oscillator. Events on a seismogram can be read to a millisecond.

7.3.2 Field procedure

When the area of investigations has been decided upon the first step in a seismic survey is to select locations for shot-holes. If only shallow holes are needed they can be prepared by augering, but generally drilling by machine is necessary in deep exploration since the shot-holes may be as much as a hundred or more metres deep. The depth of holes and the weight of dynamite for the shot (a few grammes to several kilogrammes) are important factors controlling the quality of the seismogram record.

The next step is to plant the geophones firmly on the ground which may sometimes entail burying them below the surface. In most of the work the geophones are placed along a straight line (called a profile) through the point on the surface vertically above the shot (epicentre). This procedure is known as profile shooting, but other shooting and geophone patterns adapted to particular problems are also used (Section 7.8).

The geophones are connected to the recording equipment by long cables, the amplifier gains and filters are appropriately set and when the recording film is set in motion the shot is fired. The film is stopped after a few seconds when the ground motion has substantially subsided. The moment of the shot is registered on the photographic film as a break in a continuous trace along one edge of the film.

7.3.3 Magnetic recording

A major development in seismic instrumentation is the use of the magnetic tape instead of the photographic film as the recording medium, the galvanometers being replaced by recording heads, one for each trace. The variations of the geophone output can be recorded as corresponding variations of the magnetization intensity of the tape (analogue recording). Alternatively, the output voltage can be sampled at regular intervals, say 2 ms, and converted by an analogue/digital converter to a series of numbers that are transcribed on a digital magnetic tape (digital recording). In the latter case it is not necessary to have one recording head for each geophone since a multiplexing device (high-speed electronic switch) can be employed to connect each geophone amplifier in turn to the analogue/digital converter and the recording system, for a time that is kept shorter than the sampling interval. (In truth, it is the overall multiplexing speed that limits the sampling speed). Details of magnetic recording will be found in an excellent book by Evenden and Stone [181].

One of the greatest advantages of the magnetic tape is that it allows a broad-band recording in which only the low frequency surface waves need be cut off. Considerably more information is therefore stored in such a record than in the conventional filtered seismograms. If it is desired to cut off other frequencies the tape can be played back with a suitable filter and the result recorded on a paper. Moreover, outputs from different geophones can be mixed together in playback in any desired combination. This, to some extent, achieves the same object as multiple geophone arrays to be described later. It is also possible during playback to insert weathering and elevation corrections to each of the traces by separate input units and, using suitable machines, a completely automatic data processing can be arranged.

With the development of magnetic tape recording and automatic processing, optical recording may now be said to be largely obsolete except for refraction seismic work. However, the final displays of reflection seismograms for interpretation are always analogue visual ones obtained after processing the tape data.

7.4 The refraction method

7.4.1 Parallel interfaces

The basis of the refraction method is the extension of Snell's law in optics to seismic waves. If a layer in which the waves have a velocity V_1 is underlain by another layer with velocity V_2 then by Snell's law

$$V_1/V_2 = \sin i_1/\sin i_2 \tag{7.4}$$

where i_1 and i_2 are the angles of incidence and refraction for the

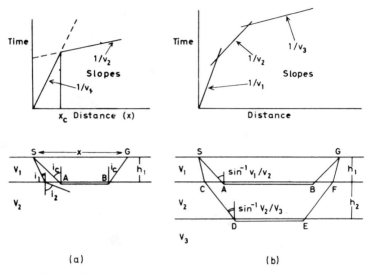

Fig. 75. Time–distance graphs in the refraction seismic method.

seismic ray (Fig. 75). The ray SA with the *critical angle* of incidence $i_c = \sin^{-1} (V_1/V_2)$ is refracted so that $i_2 = 90°$ and travels along the boundary between the two media. Obviously, this is possible only if $V_2 > V_1$.

Due to this ray the interface is subjected to oscillatory stress and each point on it sends out secondary waves and rays such as BG emerge in the top layer at the angle i_c to reach the geophone G.

If G is near to the shot S (which is assumed to be on the surface) the first 'kick' of the galvanometer will be due to the arrival of the direct wave along SG. After some time a second kick will be observed corresponding to the arrival of the refracted wave. However, if SG is sufficiently great the first arrival will correspond to the wave SABG which will have overtaken the direct wave because of a higher velocity along the path AB.

The travel time for the direct wave is $t = x/V_1$ and its plot against x will obviously be a straight line through the origin, that is the coordinate of the shot-point, with the slope $1/V_1$ (Fig. 75). Remembering that $\sin i_c = V_1/V_2$ it is easily proved that the travel time equation for the ray SABG is

$$t = \frac{x}{V_2} + \frac{2h_1 (V_2^2 - V_1^2)^{1/2}}{V_1 V_2} \tag{7.5}$$

which is also a straight line but with a slope $1/V_2$ and an intercept on the t-axis given by the second term in (7.5). This term is also called the delay-time (t_1) and can also be written as $2h_1 \cos i_c/V_1$.

Equating (7.5) with x/V_1 gives the distance coordinate at which the two straight lines intersect. Beyond this critical distance (x_c) the refracted wave arrives first at the detector. It is easy to show that

$$x_c = 2h_1 \sqrt{\left(\frac{V_2 + V_1}{V_2 - V_1} \right)} \qquad (7.6)$$

Thus a time–distance graph of the first arrivals at geophones planted at various distances from the shot will show two intersecting straight lines whose slopes give V_1 and V_2. The point of intersection x_c yields the thickness h_1 of the top layer. Alternatively h_1 may also be determined from the intercept (t_1) on the t-axis (7.5).

Clearly, $h_1 = V_1 t_1/(2 \cos i_c) = V_2 t_1/(2 \cot i_c)$. Fig. 76 shows a typical recording with the shot instant, 12 geophone traces, the 'kicks' and the times lines (10 ms apart).

For three layers with velocities V_1, V_2, V_3 $(V_3 > V_2 > V_1)$ there will be two critically refracted rays, SABG along the first

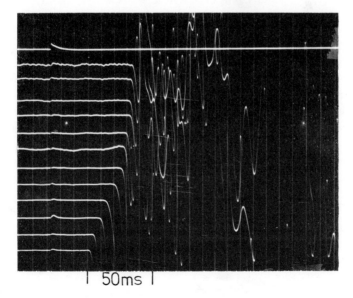

| 50ms |

Fig. 76. Refraction seismic traces on a photographic record showing shot instant (in uppermost trace) and kicks due to first arrivals.

interface and SCDEFG along the second one (Fig. 75). As before, at short distances the direct ray SG will arrive first. As the distance increases the ray SABG will overtake SG and arrive first while at still greater distances the first kick of the galvanometer will signify the arrival of the ray SCDEFG. The time–distance plot then consists of three intersecting straight segments with slopes $1/V_1$ $1/V_2$ and $1/V_3$. The thickness h_1 of the top layer is calculated from (7.6) and it can be shown that the thickness of the second layer is given by

$$x'(V_3 - V_2) = \frac{2h_1}{V_1} [V_2\sqrt{(V_3^2 - V_1^2)} - V_3\sqrt{(V_2^2 - V_1^2)}]$$
$$+ 2h_2\sqrt{(V_3^2 - V_2^2)} \tag{7.7}$$

where x' is the second intersection point.

In terms of the intercepts t_1, t_2 of the second and third segments we have the following formulae:

$$h_1 = V_1 t_1 / 2 \cos i_1$$
$$h_2 = \frac{V_2 \{t_2 - t_1 (1 + \cos 2i_3)/(2 \cos i_3 \, \cos i_1)\}}{2 \cos i_2}$$

where $i_1 = \sin^{-1}(V_1/V_2)$, $i_2 = \sin^{-1}(V_2/V_3)$ and $i_3 = \sin^{-1}(V_1/V_3)$.

These relations can be extended to any number of horizontal layers, one below the other, provided the velocity in each layer is greater than that in the layer immediately above. In general there will then be as many distinct segments on the time–distance curve as there are layers. If, however, a layer is not sufficiently thick or does not have a sufficient velocity contrast with the adjacent layers, the segment corresponding to it may be missing on the time–distance graph of the *first* arrivals. This obviously introduces errors in the depths to the lower interfaces. Such thin layers can however be detected sometimes by recording the later arrivals.

If the velocity increases continuously with the depth, for instance approximately linearly as is often the case, the time–distance graph will be a smooth curve concave towards the x-axis.

If a layer has a *lower* velocity than the one on top of it there cannot be any critically refracted ray because i_2 will always be less than i_1 in Fig. 75. No energy can be transported along such an interface and no segment corresponding to the lower layer will appear in the time–distance curve. It is worth noting, however, that such layers *could* be detected if account is also taken of shear waves.

7.4.2 Non-parallel interfaces

In Fig. 77 is shown a case where the interface between two layers is dipping at an angle with the horizontal. It is readily shown that an 'up-dip' ray such as $S_1 A_1 B_1 G$ originating at S_1 takes a time

$$t = \frac{2z_1 \cos i_c}{V_1} + \frac{x}{V_1} \sin(i_c - \theta) \qquad (7.8)$$

to arrive at the geophone while for a 'down-dip' ray $S_2 A_2 B_2 G$ from a shot at S_2,

$$t = \frac{2z_2 \cos i_c}{V_1} + \frac{x}{V_1} \sin(\theta + i_c) \qquad (7.9)$$

These relations, it may be noted in passing, will be interchanged if the surface is sloping and the interface horizontal.

The time–distance curve for the direct ray has the slope $1/V_1$ whether the ray comes from the shot S_1 or shot S_2. However, the segment corresponding to the refracted ray has a slope $\sin(i_c - \theta)/V_1$ when shooting up-dip but $\sin(\theta + i_c)/V_1$ when shooting down-dip. The reciprocals of these slopes are the up-dip and down-dip velocities, V_u and V_d respectively, and it is immediately seen that

$$\theta = \tfrac{1}{2}(\sin^{-1} V_1/V_d - \sin^{-1} V_1/V_u) \qquad (7.10)$$

and

$$i_c = \tfrac{1}{2}(\sin^{-1} V_1/V_d + \sin^{-1} V_1/V_u) \qquad (7.11)$$

(Both V_u and V_d are reckoned positive.)

The dip is directly determined from (7.10) while the depths z_1

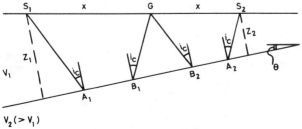

Fig. 77. Dipping interface.

and z_2 which complete the information about the configuration in Fig. 77 are obtained from the intercepts of the lines represented by (7.8) and (7.9) on the t-axis after substituting for i_c from (7.11).

The up-dip velocity $V_u = V_1 /\sin(i_c - \theta)$ is positive if $\theta < i_c$ and negative if $\theta > i_c$. If $\theta = i_c$ it becomes infinite so that the corresponding time–distance segment is horizontal. Of course, the infinite velocity is only apparent; no energy is transmitted at this velocity.

The case of an arbitrary number of dipping, but not necessarily parallel, interfaces has been dealt with in an interesting manner by Mota [182] and by Johnson [183]. Nomograms for the rapid determination of interface depths and dips have been published by several authors [184, 185].

The refraction method was much used before 1930 for oil prospecting but has now been largely replaced by the reflection method. In recent years, however, refraction work has found increasing use in civil engineering projects for bedrock investigations in connection with dam sites and hydro-electric power stations, in place of conventional drilling.

7.4.3 Example

An example of a refraction profile for a hydro-electric project in northern Sweden is shown in Fig. 78 [186]. Each dot or cross represents the travel time to a geophone placed vertically below it on the ground surface. The depths sounded on this profile are relatively shallow but the results nevertheless illustrate a number of points discussed above.

The steep initial segments near each shot-point correspond to the upper layers and at large distances from the shots these are replaced by less steep segments. This is most clearly indicated by the left-hand geophone set-up from shot 4. The break-points between segments can be determined after appropriately removing or smoothing the irregularities (such as I) caused by local inhomogeneities in the superficial layer. They may, on the other hand, be due to local variations in interface dips, for which case it is useful to remember that they are, qualitatively speaking, mirror images of an interface undulation. It should also be noted that very small dip variations can cause very pronounced irregularities of this type in the time–distance curves.

The apparent velocity in the bedrock when the geophones occupy positions around G is less when the shot is fired from S_1

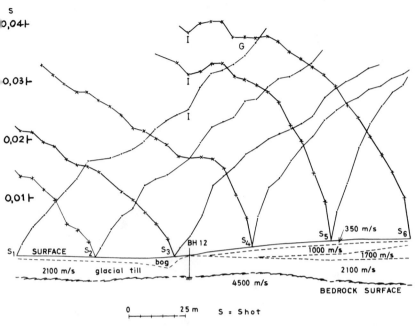

Fig. 78. Time–distance graphs and interpretation in a shallow refraction seismic survey [186, p. 269].

(steeper time–distance segment) than when it is fired from S_6 (almost horizontal segment). Evidently the ground surface and the bedrock are not parallel to each other in this region.

In the right-hand part of the figure the time–distance curves are more or less smooth indicating a fairly continuous downward increase in velocity. Finally, the reader will find an article by Green [187] on the refraction method to be a useful complement to the discussion in this chapter.

7.5 The reflection method

7.5.1 Principles

The depth to an interface between two rock formations can also be determined by measuring the travel time of a seismic wave generated at the surface and reflected back from the interface. The energy of P as well as S waves is reflected partly as P and partly as S waves. If the reflected and the incident waves are of the same kind (both P or both S) the ordinary law of reflection applies, namely, angle of incidence = angle of reflection.

It is generally assumed that the observed reflections are only
P–P reflections. This assumption is justified as a rule since near
the shot-point most of the explosion energy is transmitted as P
waves. In the reflection method the mutual separation of the
geophones is small (e.g. 25 m) compared with the depths (as much
as 5000 m) to be sounded. The geophones are generally spread
symmetrically on either side of the shot along a straight line
through it and the maximum shot–detector distance is of the
order of or smaller than the depth to the shallowest horizon of
interest. This arrangement ensures that the observed galvanometer
kicks are due to the arrival of reflected and not refracted rays.

It is easily deduced from Fig. 79a that if V is the uniform
velocity above the reflecting horizon the reflected wave arrives at
G after a time

$$t = \frac{2}{V}\sqrt{(h^2 + x^2/4)} \tag{7.12}$$

so that

$$h = \tfrac{1}{2}\sqrt{(V^2 t^2 - x^2)} \tag{7.13}$$

Either equation shows that the x–t curve is a hyperbola convex
towards the x-axis and with the t-axis as the axis of symmetry.
The line through $x = 0$, $t = 0$ with a slope $1/V$ is its asymptote.
This interesting relation is, however, of limited use in practice
since the only portion of the hyperbola that is normally observed
is the approximately horizontal one near $x = 0$.

It should be realized that if x exceeds a certain distance x_m the
angle of incidence at A will exceed i_c (critical angle of last section)
and a refracted as well as a reflected wave will arrive at G. In order
to avoid such a mixing we should keep $x < x_m$. It is easily shown
that for the simple situation in Fig. 79a

$$x_m = h \frac{V_1/V_2}{\{1 - (V_1/V_2)^2\}^{1/2}}$$

If there are two reflecting horizons which separate layers with
different velocities it is generally the practice to disregard the
refraction of rays. A ray such as SABCG (Fig. 79b) is replaced by
the ray SA′BC′G. This is justified inasmuch as the rays may be
considered to be almost vertical on account of the small
shot–detector separation.

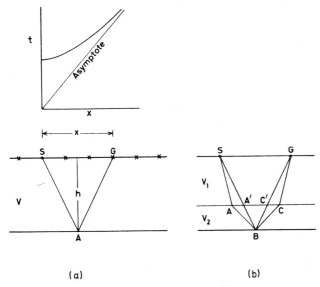

Fig. 79. Reflection seismic method.

The downward travel time of a strictly vertical ray is

$$t = h_1/V_1 + h_2/V_2 = (h_1 + h_2)/\bar{V} \tag{7.14}$$

where $\bar{V}$ is defined as the average velocity. If there are several layers the average velocity of the wave reflected from the nth horizon is

$$\bar{V} = \sum_1^n h_j / \sum_1^n (h_j/V_j) \tag{7.15}$$

and this may be substituted for V in Equation (7.13) to obtain the depth to this horizon.

Corrections for refraction can be applied if desired. Such corrections have been calculated by Krey [188, 189], Lorenz [190] and Baumgarte [191]. Several procedures can be employed for the determination of $\bar{V}$ which must, of course be known before the depths can be calculated.

One procedure is to use refraction shooting. This however requires long refraction profiles if deep horizons are to be reached, and adds considerably to the cost of the geophysical survey. It appears moreover that the velocities obtained do not always agree with the known velocities in reflection seismics. The difference may be due to the fact that the refraction velocities refer to waves travelling parallel to the strata while the reflection velocities refer to waves travelling perpendicular to the strata. If deep boreholes or wells are available in the area well-logging (Section 7.11) would be an obvious method for velocity determinations. The velocities thus obtained are, strictly speaking, valid only for the immediate vicinity of the well.

A third method employs Equation (7.12). If the geophones are well separated it is possible to determine accurately the time differences ('step-outs') between the arrivals of reflections of one and the same order at each of the geophones. If the squares of arrival times are plotted against the squares of the geophone distances from the shot the points will lie on a straight line with a slope $1/V^2$, where V is the average velocity down to the interface from which the reflection has come. Particularly reliable results are obtained with this method if the topographic relief is minimum, the weathering in the top layer is uniform and the strata are substantially horizontal, at least over the distance occupied by the geophone spread.

7.5.2 The reflection record and displays

The seismic record (whether on a photographic paper or on a magnetic tape) consists of wiggles. Part of a typical reflection seismogram recorded on a photographic paper is shown in Fig. 80. A wiggle on a single trace may represent a reflected pulse or ground movement due to noise and there would be no means of distinguishing a signal if only one trace were recorded. However, noise pulses are unlikely to be exactly in phase at all the geophones, whereas a pulse reflected from a lithologic interface

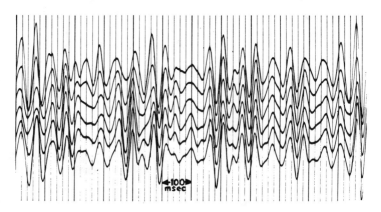

Fig. 80. Reflection seismogram (British Petroleum Company Ltd., London).

arrives at the various geophones approximately at the same time, as the geophones are close to each other. Therefore, the reflection signals are recognized by the lining-up of the crests and troughs on the adjacent traces across the entire record. The lining-up is almost straight for the later reflections, but for the earlier reflections the pulses may lie on a slightly curved line which is actually the hyperbola represented by (7.13). As a first step in the interpretation, the step-outs or, as they are also called, the normal move-outs (n.m.o.s) of the pulses, are corrected from a knowledge of $\bar{V}$. If after this correction the pulses continue to show step-outs in the arrival times the reason can be sought in a dipping reflecting horizon.

It should be remembered that a ray can be multiply reflected between two or more interfaces before arriving at a geophone, so that although all linings-up on a seismogram indicate distinct reflection events, their number need not correspond to as many actual reflecting horizons.

Fig. 81 shows another display of a seismic record. Here one side of the wiggle trace is blacked-in ('variable-area' display) resulting in considerable clarity and ease in picking out the reflections. Another visual display is the 'variable-density' type in which the photographic density along a trace is varied from black to white in proportion to signal variations. There are also hybrid systems.

Sometimes it is possible to follow reflections not only across a single record but also across the record from the adjoining geophone set-up along the shooting profile. A variable-area display of this type, constructed by juxtaposing records from consecutive geophone set-ups, is shown in Fig. 82. The scale on the vertical axis in such displays is often in seconds rather than in depth.

Fig. 81. Variable-area (VAR) display.

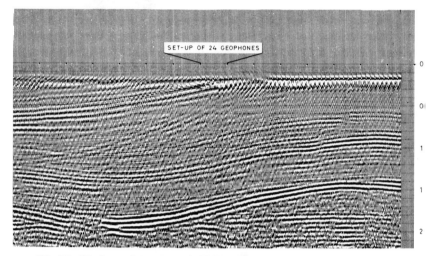

Fig. 82. Display of juxtaposed VAR records from consecutive geophone set-ups (Seismograph Service Ltd., Holwood, England, and Amoco/ Gas Council North Sea Group).

When reflections have been identified on a seismogram and the values of $\bar{V}$ are known the rest of the interpretation according to ray theory is more or less straightforward. The reflector depths are read off some convenient nomogram based on Equation (7.13)

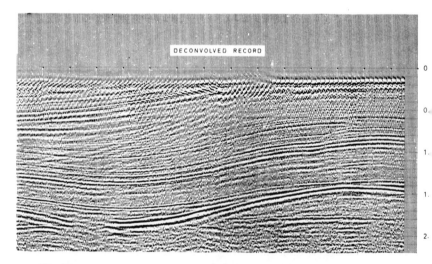

Fig. 83. Section of Fig. 82 after deconvolution (Seismograph Service Ltd., Holwood, England, and Amoco/Gas Council North Sea Group).

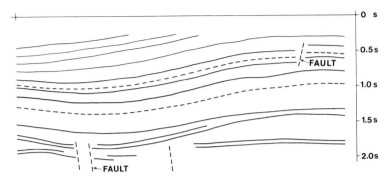

Fig. 84. Geologic interpretation (by Dr A. A. Fitch) of the section in Figs. 82 and 83.

and plotted midway between the shot and the detector on a diagram representing the vertical cross-section through the shooting profile. Certain corrections (Section 7.6) must however be applied to the observed travel times.

In another method the reflection times scaled off the seismogram are directly plotted on the cross-section map. The depths are obtained by converting the times by means of the distribution of V applicable to the area.

Fig. 84 shows the geologic interpretation of the data in Fig. 82. It traces the course of bedding planes and shows parts of two salt pillows between about 1.4 and 1.8 s. The peripheral sink between the pillows is also apparent. Faults as indicated by the seismic sections are also marked. Sections as in Fig. 82 in which the vertical axis is a time scale are called *time sections*. These can be converted to *depth sections* on the basis of known or presumed $\bar{V}$ values.

The processing of reflection seismic records is now a far more sophisticated technique than the above paragraphs would indicate. A good modern account of it is due to Geldhart and Sheriff [192].

7.5.3 Dipping interface

It has already been mentioned that if after corrections for normal move-outs, a reflection is not aligned parallel to the time lines in a display such as Fig. 80, a dipping interface may be the cause. For a plane, dipping interface the reflections line up after n.m.o. corrections along a straight line inclined to the time lines.

A graphical method known as the 'image method' is often used to determine the position of a dipping horizon. It can, of course,

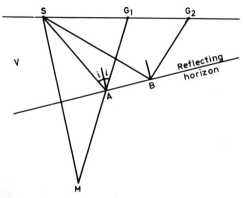

Fig. 85. Reflection at a dipping horizon.

be applied to horizontal strata as well. In Fig. 85, SAG_1 is an actual ray path and AM is the extension of G_1A equal to SA. By elementary geometry we see that M is the mirror image of S in the reflecting horizon. If t_1, t_2 ... are the travel times along SAG_1, $SBG_2, \ldots$, etc. the arcs of radii $Vt_1, Vt_2, \ldots$, etc. with the geophones as centres will intersect each other at M. Obviously the perpendicular bisector of SM represents the reflecting horizon.

The reader may find it interesting to show by using the cosine theorem of plane geometry on the triangles SMG_1 and SMG_2 that the perpendicular distance from S to the reflecting horizon is

$$d = \frac{1}{2}\left[x_1 x_2 + \frac{V^2(x_2 t_1^2 - x_1 t_2^2)}{x_2 - x_1} \right]^{1/2} \tag{7.16a}$$

where $x_1 = SG_1$, $x_2 = SG_2$.

From the same theorem it can also be shown that the dip θ is given by

$$\sin \theta = \frac{V^2(t_2^2 - t_1^2)}{4d(x_2 - x_1)} - \frac{x_2 + x_1}{4d} \tag{7.16b}$$

If G_1, G_2 are on opposite sides of S (i.e., $x_2 = -x_1$) and if we put both t_2, t_1 as approximately equal to $2d/V$, the time for perpendicular reflection, Equation (7.16b) can be immediately written in the very simple form

$$\sin \theta \approx \frac{V \Delta t}{\Delta x} \tag{7.16c}$$

where $\Delta t = t_2 - t_1$ and Δx is the geophone separation.

The position of a dipping reflecting horizon is sometimes plotted vertically below S at the calculated distance d, instead of along the perpendicular from S to the dipping horizon. Depth sections constructed in this manner are said to be *unmigrated*. In contrast, *migration* involves plotting the position in the correct direction from S. The picking-out as well as the migration of seismic events can be carried out automatically [192].

7.6 Corrections to arrival times

It is usually necessary to apply two corrections to the observed travel times of seismic waves, one for the elevation differences and the other for weathering.

The theory assumes that the shot and the detector are on the same level, but in general, their elevations differ. It is customary to reduce the arrival times by projecting the shot and detector onto a common horizontal datum plane. In refraction the actual path is then replaced by the dotted path in Fig. 86. The time difference between these paths is $(SC/V_1 - AA'/V_2)$ and is to be subtracted from the observed travel time. If the elevation of the shot is h above the datum plane $SC = h/\cos i_c$ and $AA' = S'C = h \tan i_c$. A corresponding time difference originating at the detector end must also be subtracted.

In the reflection method the paths of the waves from the shot to the detector are practically vertical and the correction is simply equal to the shot–detector elevation difference divided by the velocity or, to be accurate, the near-surface velocity of the waves.

The weathering correction arises because the weathered low-velocity layer near the surface of the earth is not homogeneous. The variations in the delay suffered by seismic waves in this layer may easily be interpreted as the variations in the depth to some deep interface between rock formations.

The corrections are usually made by determining the actual

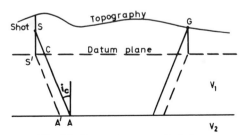

Fig. 86. Corrections to arrival times.

thickness of the weathered layer by a series of shots with small charges and close shot–detector spacings, but rapid graphical methods can also be devised, especially in shallow refraction work. The details of the correction procedures vary considerably. Some of them have been described by Dobrin [180]. A theoretical treatment of the effect of the weathered layer will be found in a paper by Menzel and Rosenbach [193].

7.7 The seismic pulse

7.7.1 Propagation of a pulse

The simple ray theory outlined above is adequate for most purposes in applied seismology, but it does not give a complete physical description of seismic phenomena. In recent years, however, a great body of literature has grown around the physical theory of the propagation of elastic waves in stratified media and, following the lead of Ricker [175] particular attention has been focused on the actual seismic impulse initiated by an explosion [194].

When a charge is detonated the material around it is permanently distorted, a cavity is formed and an outward-travelling seismic impulse (wavelet) originates. Even at distances as large as a couple of metres from the charge the maximum stress experienced by the material may exceed the range of reversible stress–strain relationships and a permanent 'set' may be evident. The distance around the shot at which the maximum stress falls just within such relationship defines the *equivalent cavity*. This region rather than the shot itself is often regarded as the source of the seismic pulse.

According to Ricker's theory, the centre of the seismic pulse travels with a velocity characteristic of the medium, and the breadth of the pulse (measured in μs) increases with increasing propagation time according to some definite law. In shales, for example, the pulse breadth is proportional to the square root of the time.

If the pulse encounters a boundary between two geologic formations with different acoustic impedances (product of the velocity of elastic waves and the density of a formation) a reflection will occur and a broadened pulse will be received at the earth's surface after some time. This is illustrated qualitatively in Fig. 87, after Anstey [195]. The form and breadth of the pulse and hence its frequency spectrum will depend on the travel time.

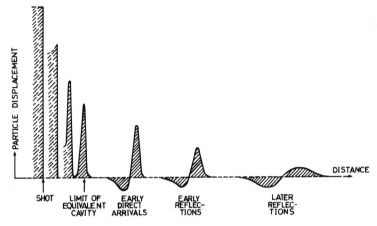

Fig. 87. Broadening of a seismic pulse.

The pulse will also be distorted further to a greater or lesser extent by the recording equipment.

In addition to some large velocity contrasts which cause strong reflections there will, in general, be a great number of minor velocity contrasts in a geological column. Each of these as well as other inhomogeneities will initiate a reflected pulse which arrives at the surface in broadened form. The seismogram record between two strong reflections consists not merely of 'noise' but, to a large extent, of such reflections superimposed on and interfering with each other. What remains after these *signals* are subtracted would be the noise due to surface and transverse waves, multiply reflected waves, transmissions through the weathered layer and wind.

It will now be appreciated that the seismic record contains a great deal of information about sub-surface inhomogeneities but that only a small part of it can be extracted by the simple ray theory.

7.7.2 Attenuation of a pulse

Seismic waves are reduced in amplitude as they are propagated through the earth due to three factors: geometrical divergence, partial transmission and reflection at acoustic boundaries and absorption of energy in the medium of transmission.

The influence of the first factor is well known and can be allowed for, e.g. at large distances from the source spherical waves reduce in amplitude in inverse proportion to the distance travelled.

The transmission and reflection coefficients of a geological interface are functions of the elastic contrast between the layers in contact. In principle, a comparison of the amplitudes of the reflected pulses should provide some information about such contrasts. Although the theory of this topic has been well developed [196, 197] the widespread use of automatic gain control in seismic amplifiers has until recently virtually prohibited this comparison. With modern digital recording systems with their wide dynamic range the subject is likely to attract more attention in the future.

The attenuation of plane waves due to absorption of energy is of the familiar exponential form: $A = A_0 \exp(-\delta f x/V)$ where δ is the logarithmic decrement, f the frequency, x the distance and V the velocity. Typical values of δ for the earth materials in bulk would be around $0.02-0.03$.

Two absorption mechanisms, viscosity and solid friction, have been suggested. Both mechanisms are observed in rocks but it seems that solid friction generally predominates. The presence of water apparently increases the decrement and at the same time leads to a predominance of viscous damping.

Work on the attenuation of seismic waves has been reported by Born [198] and, more recently, by Datta [199].

7.8 Filtering and geophone arrays

7.8.1 Filtering

The object of filtering is to exclude the noise referred to earlier. This noise, which is mainly due to surface waves ('ground roll') and wind, is liable to mask the reflection signals. The common practice in filtering is to cut off the low frequency ground roll by a high-pass filter and the high frequency wind noise by a low-pass filter.

Filters have a disadvantage, however, in that they necessarily distort the shape of the signal pulse. In particular the pulse is lengthened and moreover its phase is shifted, that is, characteristics such as crests and troughs are displaced in time as illustrated in Fig. 88 for a pulse that initially has a shape of a step. The result is that filters create interference patterns on the seismogram in addition to those created by the signals.

Nowadays filtering of a numerical rather than instrumental nature is also common. In this the seismic records are 'operated upon' so as 'to reverse the effect of the transmission medium and

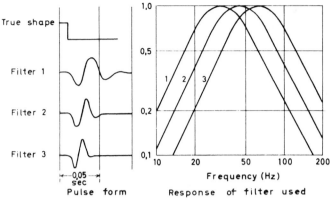

Fig. 88. Effect of a filter on a step signal.

contract and narrow the wavelet to make up for the broadening it underwent in transmission'. The object is to derive the original shape. This procedure, known as deconvolution, is described in Section 7.10. Instrumental as well as numerical filtering has been comprehensively discussed by Smith [200].

7.8.2 Multiple geophone arrays and stacking

Besides filtering, another method is also used to eliminate or minimize the noise in seismic records. The geophones are laid out along the profile with very close separations (e.g. 10 m) and the outputs of several adjacent geophones are added together and recorded as a single trace.

Consider a surface wave of some definite wavelength. If the geometry of the geophone groups is appropriately selected the instantaneous motion of the ground due to the vertical component of this surface wave will be upwards at some geophones and downwards at others, so that the sum of the outputs will eliminate the surface wave from the record.

Actually, the effect of multiple arrays is more complicated than the brief description above would indicate. For one thing, it is unrealistic to consider only one isolated wavelength because seismic signals and noise in practice contain a continuous spectrum of waves. Broadly speaking, however, it can be shown that if n inputs containing the same signal but random noise are added together the root-mean-square signal in the output is increased by a factor n (provided the n signals are coherent) whereas the r.m.s. noise is increased by a factor $\sqrt{n}$. Hence the signal-to-noise ratio in the output is improved by the factor $\sqrt{n}$.

The summation of signals from a number of similar input channels for the purpose of increasing the signal-to-noise ratio is known as *stacking*. There is a large variety of stacking layouts but we shall consider only two of these.

In *vertical stacking* we have one shot-point and a geophone, both of which are kept at fixed locations. The outputs of the geophone for a number of shots are added, either in the field or later, in the digital processing of data. The origin of the term vertical stacking is understood if we imagine each wiggly record to be placed below the previous one in the correct time relation and a sum taken down the entire ensemble.

Common depth point (CDP) or, as it is also called common reflection point (CRP), stacking was originally suggested by Mayne [201] to enhance weak reflections in relation to background noise, but it has also proved to be useful for reducing the signals from multiple reflections (the so-called signal-generated noise). The principle of CDP stacking can be seen from Fig. 89 where the numbers represent the traces on a record and A,B,C,D,E,F are 6 successive shots. For each shot we record traces from 24 geophones along a line. Actually each trace usually represents the signal from an array of geophones at a location and not just a single geophone. Also, for convenience, the geophone set-ups for the shots are shown in Fig. 89 as displaced from each other. In practice, the line is one and the same.

Assuming a horizontal reflector we see that trace 21 on the record for shot A has the same reflection point as trace 17 for shot B, trace 13 for shot C, etc. The traces 21,17,13,9,5 and 1 for shots A,B,C,D,E and F respectively are added together during the

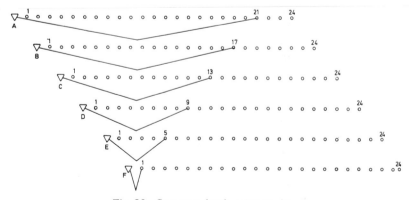

Fig. 89. Common depth point stacking.

processing to enhance the signal from their common reflection point. Similarly it can be seen that traces A-24, B-20, C-16, D-12, E-8 and F-4 have the same reflection point and can therefore be added together*.

The arrangement of Fig. 89 gives a 6-fold stack for each depth point. Other arrangements with more numerous geophones will of course give a higher stacking multiplicity. Although the CDP procedure implies a considerable degree of redundancy most reflection seismic surveys now use it routinely.

7.9 Convolution and synthetic seismograms

Seismic reflections arise from changes in the acoustic impedance of the ground. If the reflection coefficient $(V_2 d_2 - V_1 d_1)/(V_2 d_2 + V_1 d_1)$ where d is the density, were known at each boundary it would be possible to calculate the amplitude of a reflected pulse, and by combining the various pulses in the correct time relation we could 'synthesize' a seismogram record.

If the coefficient is positive, an impinging compression in a pulse will be reflected as a compression and a dilatation as a dilatation, while a negative coefficient causes a compression to be reflected as a dilatation and vice versa. Moreover, in either case the displacement in the pulse is reduced in direct proportion to the reflection coefficient.

In Fig. 90a is shown a wavelet representing the ground disturbance at the surface when a pulse reflected from an interface with an ideal reflection coefficient 1 arrives at the surface. It will be sufficiently accurate for our purpose to characterize this wavelet by a series of uniformly spaced ordinates 0, −4, −12, −8, 40, 16, −24, −8, 0 in arbitrary units. On reflection from an interface 1 having a reflection coefficient 0.5, the arriving pulse will produce half as much disturbance (assuming that the ground is a linear medium and the pulse undergoes no further change of shape in reaching the surface).

The ground disturbance due to a reflection from another interface, 2, one-half time unit 'deeper' than interface 1 (that is, $V/2$ depth units apart from 1) will start one time unit later (since the outward going 'Ricker pulse' has to travel to and back from interface 2 in being reflected from it). Suppose the reflection coefficient of interface 2 is 0.25. Then a series of ordinates 0, −1, −3, −2, 10, 4, −6, −2, 0 displaced one time unit in relation to the

---

*To avoid confusion the reader should note that some descriptions of the CDP method use geophone numbers rather than trace numbers.

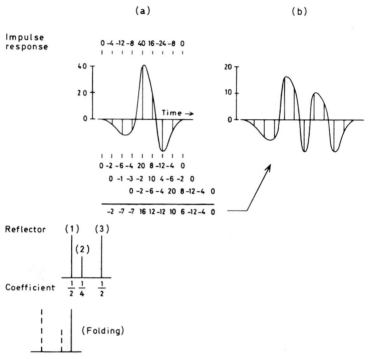

Fig. 90(a and b). Superposition and convolution.

earlier series 0, −2, −6 . . . will represent the ground motion due to this reflection. Similarly a third reflector one time unit deeper than interface 2 and having, say, a reflection coefficient 0.5, will produce a disturbance displaced two units in time. Adding (i.e. superposing) the disturbances we see that the resultant ground disturbance will be represented by the pulse in Fig. 90b with sampled ordinates 0, −2, −7, −7, 16, 12, −12, 10, 6, −12, −4, 0 (or more stringently expressed, by the ordinates series which we may use to reconstruct a first approximation of the pulse form).

Exactly the same result can be obtained if, instead of superposition, we use the following procedure. The string of reflection coefficients (1), (2) and (3) is folded back on itself (convolved, cf. the aptly named genus of plants *convolvulus*) and slid past in discrete steps of one time unit, across the series 0, −4, −12, . . . etc. For each position of the string we multiply each of the three reflection coefficients by the pulse ordinate of this series found directly above that coefficient, and then add the three results together. We obtain again the series 0, −2, −7, . . . This procedure is known as convolution and, as seen, its outcome is exactly the same as that of superposition.

The final appearance of a seismic record is, of course, also influenced by multiple reflections, noise and the change of pulse in transmission but, in essence, the record is the convolution of a basic Ricker wavelet with the long string ('log') of all the various reflection coefficients in the ground. If the log is known a synthetic seismogram can be constructed by convolving the log with the basic pulse.

Ideally, a synthetic seismogram will exactly duplicate the corresponding noise-free field record. However, such duplication would be of no more than academic interest but for one fact. It is that synthetic seismograms furnish the possibility for studying the changes produced in the seismic record by changes in the assumed ground characteristics. This study in turn may enable the interpreter to identify particular geological sections. Furthermore, seismograms can be calculated for points other than the actual geophone plants. Such calculations would indicate any advantage that could accrue from additional recordings in the area and the changes one must look for in the records in that case.

The first paper on synthetic seismograms was published by Peterson *et al.* [202] and since then the use of the technique has been increasing. For details the interested reader may be referred to a symposium on the subject [203].

7.10 Deconvolution

It is evident from Fig. 90 that the effect of convolving the seismic pulse with the reflection coefficient log is to stretch or broaden the pulse. The object in seismic interpretation is the opposite, namely, to recover the original log by compressing (or deconvolving) the seismic record. The procedure is also called inverse filtering.

If s, p and l denote the seismic record, the basic pulse and the reflection log (as function of time), convolution is symbolically expressed by the equation

$$s = p * l$$

Denote by S, P, L the respective Fourier transforms. Then (Appendix 7, p. 259).

$$S = PL$$

where the right-hand side is ordinary multiplication.

Since *S*, *P* can be expressed as polynomials in a common variable, we can obtain *L* (as a polynomial) by the polynomial division *S/P*. Computer programs exist for calculating transforms as well as for polynomial divisions [204]. The desired log is given by the inverse Fourier transform of *S/P*. More generally we can write

$$L = FS \qquad (7.17)$$

where *F* is a 'filtering function'. A suitable form for *F* is, for example, $\cot(\omega t/2)\sin(m\omega t)$. If *f* is the inverse Fourier transform of *F*, the log can also be obtained by convolving *f* with *s*,

$$l = f * s \qquad (7.18)$$

In practice, deconvolution involves three principal stages:

(1). Calculation of the autocorrelation function (ACF) of a suitable section of the record, that is, estimation of the integral

$$(1/2T) \int_{-T}^{T} s(t)s(t + \tau)\, \mathrm{d}t$$

for different values of τ. The procedure is computationally similar to convolution except that the sliding trace is the same as the stationary trace (nor is it folded on itself). By a 'suitable section' is meant a section that may be regarded as a random superposition of many elementary pulses each reasonably close to some average shape. *T* denotes the length (seconds) of the section.

(2) Choice of *f*. This requires, in the first place, an estimation of the desired filter *F*. It turns out that *F* is effectively the Fourier spectrum of the central portion of the ACF. Hence the need of calculating ACF in stage (1). Having estimated *F* we get *f* as its inverse Fourier transform.

(3) Application of (7.18). Since $f(t)$ is now selected and $s(t)$ is given this final step yields the desired log *l*. The log is obtained as a function of time but can be converted to one of depth, by appropriate assumptions about the average velocity $\bar{V}$.

It will be seen that the entire deconvolution process is very well suited to operation on high-speed computers. It is now routinely used in the interpretation of seismic reflection data. Fig. 83 shows the section of Fig. 82 after deconvolution. The reflections are now more clearly visible due to pulse compression. It must not, however, be imagined that deconvolution always leads to a clearer display. The reason is that the inverse filter can act adversely on the noise wiggles and multiply reflected signals and may amplify them.

Considerable research has gone into designing optimum inverse filter functions and in devising rapid computer procedures for deconvolution. For details reference should be made to the special literature on the subject [205−208].

7.11 Continuous velocity logging (CVL)

It has already been mentioned that in order to convert seismic time sections into depth sections it is necessary to know the average velocity $\bar{V}$. If a well is available a geophone can be lowered in it and shots are fired on the surface in the vicinity of the hole. Knowing the arrival time of the seismic pulse and the shot−geophone distance $\bar{V}$ is easily calculated.

Another method is the CVL method first introduced by Vogel [209] and Summers and Broding [210]. The idea is to measure the seismic velocity over very small intervals by lowering a suitable apparatus in a borehole. The acoustic pulse is produced by a powerful electric arc discharge in a small transmitter and the signal is picked up by a receiver only about 1−2 m away in the hole.

A part of CV-log is shown in Fig. 91 together with the corresponding geologic section. The total travel time curve shown is obtained by integrating the interval times. The solid triangle at a depth of about 2330 ft indicates the position of the geophone used to calibrate the CVL. The travel time to this geophone was 0.243 s.

It is evident from the CVL that the limestones in this instance give high velocities (17 000 ft/s) while crystalline anhydrite and marls with bands of dolomite, limestone and anhydrite have generally low velocities, sometimes of the order of 7000 ft/s.

Although CVL is subject to various corrections and inaccuracies it is generally assumed that the local velocity variations evinced by it are geologically significant. The method is now part of the routine of applied seismology.

7.12 VIBROSEIS

This ingenious system of seismic exploration (a trademark of the Continental Oil Company) uses a vibratory instead of an impulse source. It is well known that an impulsive pulse introduces a definite band of frequencies in the earth (Appendix 7, p. 259). In the VIBROSEIS system the signal used to introduce this band is a

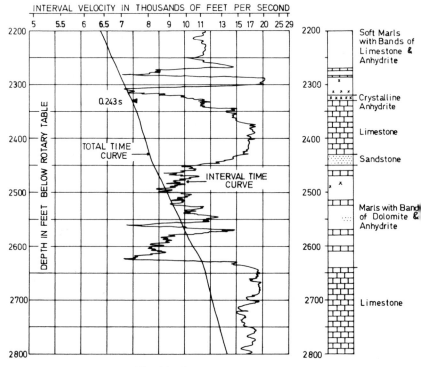

Fig. 91. Continuous velocity log.

linear frequency sweep of the form shown in Fig. 92a. In radar technology such signals are often called 'chirps' or 'pulse compressions'. Typically, the frequency is swept from 15 to 90 c/s in about 7 seconds. The signal is introduced by truck-mounted hydraulic or electromagnetic vibrators capable of exerting a total thrust of several tons on the earth. The input seismic pulse of the impulse sources to which the VIBROSEIS signal corresponds is given by the autocorrelation function (p. 218) of the signal. The centre of the pulse thus obtained (Fig. 92g) corresponds to the time break of the conventional methods.

The returning signal from each reflector in the earth is likewise a frequency-swept signal. Three such signals are shown in Figs. 92b, c and d. They add up to the recorded signal (Fig. 92e). To recover the reflected pulses, the total signal is cross-correlated with the input signal. Computationally this procedure is the same as

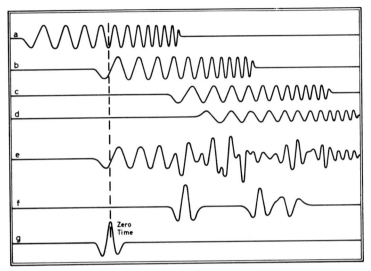

Fig. 92(a–g). Principle of VIBROSEIS.

convolution (p. 259), except that the cross-correlated trace is not folded but slid directly past the reference trace. For each position of it, the ordinates on the two traces are multiplied and summed. This leads to Fig. 92f which is nothing but a conventional seismic trace.

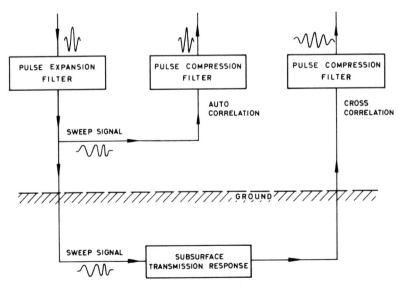

Fig. 92h. Block diagram of a VIBROSEIS system.

Theoretically, the VIBROSEIS system is equivalent to conventional seismic exploration, but differs from it in the method of recovering the seismic information. However, it has several operational and practical advantages. First, the system is inherently safer and more convenient than one using explosives. Second, the sweep rates and frequencies can be optimally selected to suit the geologic conditions in the area so that power is not wasted in generating frequencies which the earth will not transmit. Finally, the energy injected into the earth can be increased almost without limit by coupling together as many vibrators as desired and making them work in phase from a master vibrator.

A complete block-diagram of the system is shown in Fig. 92h. The reader will find further details about VIBROSEIS elsewhere [211–213].

8 Radioactivity methods

8.1 Introduction

The geophysical methods employing radioactivity came into prominence with the demand for uranium metal in atomic reactors. However, the methods are not restricted in scope only to the search after the ores of radioactive metals or minerals associated with them (e.g. placer deposits of titanium or zirconium) but can often be used with advantage in geological and structural investigations as well.

8.2 Theoretical background

The nucleus of an element X with an atomic number Z has a positive electric charge of Z (atomic) units and is made up of nucleons (protons and neutrons). The number of nucleons is the mass number A of the element and the nucleus is symbolically denoted as $_Z^A X$. Elements with the same Z but with different A are said to be *isotopes* of each other.

Certain nuclei disintegrate spontaneously emitting α-particles (helium nuclei $_2^4 He$) and β-particles (electrons and positrons). This is the phenomenon of radioactivity. These emissions alter the nuclear charge, α by -2, β^+ (positron) by -1 and β^- (electron) by $+1$. This means that the disintegrating nucleus is transformed into a nucleus of another element. Very often the daughter nucleus is also radioactive in its turn.

The nucleus is generally in an excited energy state after a β emission and returns to its ground state with the emission of a further particle, the γ-quantum or ray. In some rare instances an α emission too is followed by a γ-ray, e.g. in radium. The γ-particle is a purely electromagnetic radiation which does not alter the nuclear charge.

The disintegration of a given quantity of any radioactive

element can be expressed by the formula $N = N_0 \exp(-\lambda t)$ where N_0 is the number of nuclei initially present and N the number remaining after a time t. λ is known as the decay constant and after a time $1/\lambda$ the number of nuclei present is reduced to $1/e \approx 1/3$ of the initial number.

A given quantity of a radioactive element is halved after a time $T = \ln 2/\lambda = 0.693/\lambda$. This time is known as the *half-life* of the element.

About 50 natural and more than 800 artificial radioactive nuclei are known. Natural radioactivity is confined principally to four *radioactive series* which start from the following isotopes of neptunium, uranium and thorium: $^{239}_{93}\text{Np}$, $^{238}_{92}\text{U}$, $^{235}_{92}\text{U}$ and $^{232}_{90}\text{Th}$. The transformations of the ^{238}U series are shown in Table 13.

The uranium and thorium series end in the stable isotopes of lead $^{206}_{82}\text{Pb}$, $^{207}_{82}\text{Pb}$, $^{208}_{82}\text{Pb}$. The neptunium series ends in the bismuth isotope $^{209}_{83}\text{Bi}$.

Besides the members of the four radioactive series, at least 10 other naturally occurring isotopes, all of elements with atomic numbers less than that of lead, are known to be radioactive. Chief among these is the isotope $^{40}_{19}\text{K}$ of potassium with a half-life of 4.5×10^8 years. About 42 per cent of ^{40}K is transformed with an emission of a β-particle into calcium 40 and about 58 per cent with the capture of an electron from the K shell of potassium into

Table 13   Radioactive disintegration of the uranium-238 series

Element	Z	Emission	Half-life	Product
Uranium-238	92	α	4.51×10^9 y	^{234}Th
Thorium-234	90	β, γ	24.1 d	^{234}Pa
Protactinium-234	91	β, γ	1.14 min	^{234}U
or				
Uranium-234	92	α	2.52×10^5 y	^{230}Th
Thorium-230	90	α	80 000 y	^{226}Ra
Radium-226	88	α, γ	1622 y	^{222}Em (Radon)
Emanation-222	86	α	3.825 d	^{218}Po
Polonium-218	84	α, β	3.05 min	^{214}Pb, ^{218}At
Lead-214	82	β, γ	26.8 min	^{214}Bi
Astatine-218	85	α	2 s	^{214}Bi
Bismuth-214	83	β, α, γ	19.7 min	^{214}Po, ^{210}Tl
Polonium-214	84	α	1.6×10^{-4}	^{210}Pb
Thallium-210	81	β, γ	1.3 min	^{210}Pb
Lead-210	82	β, γ	20 y	^{210}Bi
Bismuth-210	83	β, α	5.0 d	^{210}Po, ^{206}Tl
Polonium-210	84	α	138.4 d	^{206}Pb (stable)
Thallium-206	81	β	4.2 min	^{206}Pb (stable)

argon 40. The transformation is accompanied by high energy γ-radiation of a wavelength of about 0.8 pm.

The α- and β-particles lose their energy in passing through matter by collisions, ionization, etc. and are brought to a virtual stop within a certain distance which is called their *range*. In air at 18° C the range of α-particles is only a few centimetres, in denser substances, e.g. mica or aluminium, it is still smaller being of the order 30 μm. Even the β-particles whose range is several hundred times greater are completely stopped by a thin sheet of lead or, for example, a few centimetres of sand.

The intensity of γ-rays in traversing matter decreases exponentially with distance so that we cannot speak of a definite range in this case. Theoretically, γ-rays could be detected across any thickness of matter but a practical limit is set by the sensitivity of the detecting instruments and the background effects due to cosmic radiation. For practical geophysical purposes γ-radiation may be taken to be entirely absorbed by $1-2$ m of rock.

Evidently, the α- and β-particles are not of much avail in geophysical field work since they will be indetectable as soon as a radioactive deposit has the thinnest cover of overburden. Thus, the search for radioactive minerals is, to a large extent, a search for places with abnormally high γ-radiation. Uranium, for example, is located *indirectly* from the powerful γ-radiation emitted by two products of the uranium series: $^{214}_{82}$Pb and $^{214}_{83}$Bi.

According to the quantum theory, electromagnetic radiation consists of discrete 'particles', namely photons. The energy of a photon is given by hc/λ, where h is Planck's constant, $c = 3 \times 10^8$ m/s is the velocity of light, and λ is the wavelength of the radiation. The energy is commonly expressed in electron volts (eV). One eV is the energy (1.602×10^{-19} joule) acquired by an electron in falling through a potential difference of one volt.

Since $h = 6.625 \times 10^{-34}$ joule s $= 4.14 \times 10^{-15}$ eV s, the energy of a photon is $1.24 \times 10^{-6}/\lambda$ eV. The wavelengths of γ-rays are of the order of $10^{-11}-10^{-12}$ m ($0.1-0.01$ Å) and the corresponding energies are of the order of $0.1-1$ MeV.

A radioactive series, such as that in Table 13, is said to be in equilibrium when as many atoms of any unstable element in it are being formed per second as are disintegrating. Such a series then emits a definite spectrum of γ-rays. It is customary to describe the spectrum in terms of energy levels, instead of wavelengths as for the spectrum of ordinary light. The γ-spectrum of ^{238}U contains lines at various levels from about 0.1 MeV to about 2.4 MeV. The spectrum of ^{40}K consists of a single line at 1.46 MeV. A spectral line is never perfectly sharp but has a finite, although usually small, energy width.

By measuring the energy emitted at a level it is possible to determine the amount of the particular γ-emitting nuclide present in the sample, if at least this nuclide is in radioactive equilibrium with respect to its parent and daughter. If the whole sample is in radioactive equilibrium and if energy lines from other nuclides or other radioactive series are not overlapping the measured line, the content of the parent element of the series (U, Th) can be assessed by comparison with a standard of known content. Conditions are never as ideal in measurements on rock outcrops, but postulating radioactive equilibrium the measured intensities can still be converted to U or Th content, which then is referred to as equivalent content (eU, eTh).

8.3 Radioactivity of rocks

Minute traces of radioactive minerals are present in all igneous and sedimentary rocks, in oceans, rivers and springs, in oil, and in peat and humus.

The average amounts of U, Th and ^{40}K in a few materials of the earth's crust are shown in Table 14. From this we see that there is a large difference between the radioactivity of basalts and granites. Moreover, the latter have a remarkably high content of ^{40}K. This

Table 14   Radioactive contents

	U *g/t*	*Th* *g/t*	^{40}K *%*
Basalt	0.9	4.2	0.75
Diabase	0.8	2.0	
Granite	3–5	13.0	4.4
Sediments	~1.0	5–12	
Limestone	1.3	1.1	
Oil	100		
Ocean water	0.000 15– 0.0016	0.0005	
	%	*%*	
Uraninite	38–80	0–0.3	
Carnotite	50–63(†)	0–15	
Thorianite	23–26	53–57	
Monazite	0.02–0.7	3.5–16.5	

†Contents of uranium oxide.

fact is of great consequence because granites are very common rocks and the γ-radiation from their potassium produces a radioactive background which may make it difficult to locate uranium and thorium ores. Sometimes the radioactivity of potassic feldspars in pegmatite dikes may be misinterpreted as being due to a concentration of uranium and thorium.

The high uranium but low thorium content of oil is also striking.

8.4 Radiation detectors and field procedure

8.4.1 Detectors

The α-, β-, and γ-radiations are detected by their ionizing action. In geophysical work only the γ-rays can normally be detected because the α- and β-particles are easily stopped by matter.

One common type of detector is the Geiger counter. It consists of a glass tube containing some gas (usually argon with a small amount of alcohol or amyl acetate) with cylindrical cathode round a wire anode. The electrodes are kept at a high potential difference. When a γ-ray passes through the gas it produces ions which are accelerated by the field and produce further ions. The momentary current passing through the tube can be amplified and registered on a meter or heard as 'click' in a pair of headphones.

The Geiger counters respond to only 1 per cent or less of the incident γ-rays. On the other hand they register practically all *corpuscles* in the cosmic rays.

A more efficient type of detector is the scintillation counter. This utilizes the fact that certain crystals such as thallium-activated sodium iodide, anthracene, para-terphenyl, etc. scintillate when they absorb γ-rays. The scintillations can be detected by a photocathode in a photomultiplier tube and recorded suitably after amplification.

The scintillation counters are almost 100 per cent efficient in detecting γ-rays. Their sensitivity to cosmic rays is about the same as that of Geiger counters so that their relative response to γ-rays is much higher.

Detectors are constructed in two basic modes. The *differential spectrometer* records only radiation falling within predetermined upper and lower energy limits. If the limits are very close together (say a few tens of keV at the most) the spectrometer is said to respond to a channel or a line, while for wider separation of limits (several hundred keV) it is said to respond to a window.

The *integral spectrometer* is set to exclude radiation below a predetermined energy level, the threshold, and records all radiation having photon energies greater than this level. In some instruments the threshold can be varied. There exist also mixed-mode instruments. The choice of the spectrometer will depend to a large extent on the purpose and the requirements of a survey. Window and threshold spectrometers are more rugged, more rapid and require less checking than channel spectrometers. But the latter allow precise determinations of the various nuclides. Under controlled conditions determinations within a few g/t U or Th are possible. Channel spectrometers are, however, expensive.

8.4.2 Field procedure

Radioactive surveys may be made either as 'spot examinations' or, more systematically, on a grid pattern of lines and points with the observer walking along pre-laid lines holding the detector some 40−50 cm above the ground. They can also be carried out from a car by mounting the detector on the roof provided the roads are unsurfaced, for the radioactivity of the materials in 'metalled' roads vitiates the observations.

The intensities are often recorded as counts-per-minute but most instruments are nowadays calibrated in milliröntgens-per-hour. The röntgen is the quantity of γ- (or X-) radiation which produces 2.083×10^9 pairs of ions per cm^3 of air at NTP. It corresponds to an energy absorption of 8.38×10^{-6} joules per kg of air.

The intensities recorded are the integrated effects of the radiation from a finite area on the ground. In foot-surveys most of the intensity comes from within a circle of about 3 m radius while in car-borne surveys about 90 per cent of it comes (because of the greater detector height) from within a radial distance of 15−20 m.

Topographic irregularities, absorption and scattering of radiation in the earth materials, the dispersion of radioactive materials due to weathering, etc. and the 'background radiation' are some of the factors affecting instrument readings. These must be carefully considered in the interpretation.

The background is due mainly to cosmic rays, potassium 40 and the minute quantities of uranium and thorium which are almost always present in rocks. It may vary from one area to another as well as within a single area. A reading cannot be considered signficant in general unless it is 3−4 times the background.

An example of a radioactive survey is shown in Fig. 93, after

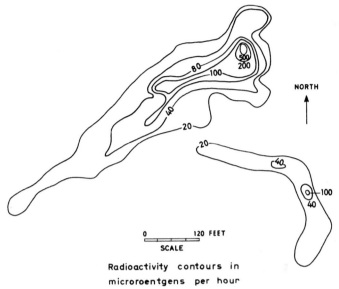

Radioactivity contours in
microroentgens per hour

Fig. 93. Radioactivity survey [214].

Moxham [214]. The area (Pumpkin Buttes area, north-eastern Wyoming, U.S.A.) is underlain by sandstone, shale and coal of Paleocene and Eocene age. Deposits of secondary uranium are found in 'rolls' and disseminated in several sandstone layers of the Eocene.

Contrary to the above example of high radioactivity over ores, Zeschke [215] has reported that 20 deposits of manganiferous ore in south-eastern Europe showed a systematic *decrease* in activity as one approached the outcrops. Above the outcrops the activity was practically nil (i.e. = background). Apparently, the ore deposits have been formed by descending solutions, and the water that flowed through the openings now occupied by ore probably leached most of the radioactive material from the surrounding marble rocks which, it is presumed, had originally a higher radioactivity than at present. This example indicates that the interpretation of radioactivity surveys, like that of all other geophysical surveys, must be done in conjunction with geological information.

8.5 Radon measurements

If radium is present at a place its disintegration product, radon gas (an α emitter with $T = 3.825$ days), may seep towards the surface with currents of water, diffuse through permeable rocks or escape through fissures and cracks.

Radon is present in suspension in nearly all well and river waters and in oil. In some cases the concentration may be very high as in the waters from Valdemorillo, Spain or Manitou, Colorado whose radon activities are 0.22 and 0.03 μc/l. (One curie is the activity corresponding to 3.7 x 10^{10} disintegrations per second = activity of 1 g radium.)

Radon measurements have been used as a geophysical method [216]. A small tube is thrust into the ground to a depth of one metre, or thereabouts. A sample of gas is drawn by means of a hand-pump into a container connected to the tube and is analysed for its α-activity. A high activity may be due to radon and may in turn indicate faults, fissures, uranium veins, etc.

8.6 Radioactive density determinations

The absorption of γ-rays has been employed in recent years for determining soil density in foundation investigations and hydrological problems. For this purpose an artificial radioactive isotope such as ^{60}Co (T = 5.26 years) or ^{137}Cs (T = 33 years) is used. The former emits γ-rays during beta disintegration, the latter changes to an excited ^{137}Ba which in turn emits γ-rays in returning to the ground state.

A metal probe with the γ-source at the end is driven vertically to a depth (z) of about 50–100 cm. The count of a detector placed on the surface at various horizontal distances (x) along a line is noted. The soil will absorb γ-rays so that the count will depend upon its density ρ.

If the detector is assumed to be very small it can be shown that

$$N = \frac{N_0}{4\pi r^2} \mu \exp(-\mu r) \tag{8.1}$$

where N_0 is the total number of γ quanta emitted, N the number reaching the detector, $r = \sqrt{(x^2 + z^2)}$, the source–receiver distance, and μ an absorption coefficient.

The plot of $\ln(r^2 N)$ against r will evidently be a straight line with a slope μ. It can be proved however that $\mu = \mu'\rho$ where μ' depends only upon the energy of the radiation. The mean values of $1/\mu'$ are 18.4 g/cm^2 for ^{60}Co and 14.2 g/cm^2 for ^{137}Cs but are affected slightly by test conditions. For the examples shown in Fig. 94 (after Homilius and Lorch) using ^{137}Cs $1/\mu'$ = 15.33 g/cm^2 (clay) and 15.65 g/cm^2 (sand).

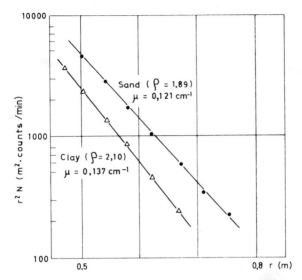

Fig. 94. Density determinations by radioactivity [217].

Density values accurate to 1 per cent or better can be obtained with this method whose chief advantage is that the density can be determined *in situ*, without having to take samples.

A rigorous and detailed discussion of the method has been given by Homilius and Lorch in two interesting papers [217, 218].

8.7 Airborne radioactivity measurements

As with the other methods, airborne radioactivity work differs from the ground work mainly in respect of the operational procedure. The general principles of measurement and interpretation are the same in either case.

The intensity of the radiation from radioactive sources such as uranium veins decreases rapidly with the height of the observation point above the earth's surface. The effect is due, in part, to geometrical divergence and, in part, to absorption and scattering in the air. Needless to say, scintillometers of the highest sensitivity are needed in airborne work and flight heights must be kept relatively low (50–70 m). The United States Geological Survey, for example, has used instruments with up to six thallium-activated NaI crystals (some ten centimetres in diameter and five centimetres thick) coupled in parallel. Such a detection system is reported to be sensitive to differences as small as 10 parts per million in the uranium content of rocks.

The detectors are generally shielded (except for the face of the crystal which is directed downward) from cosmic rays by means of a lead shield. The radiation intensities are recorded continuously and the deflections on the recording chart are calibrated by flying the detector over naturally occurring or artificially placed radioactive sources of known strength.

Considerable attention has been paid to devising methods for quantitative estimations of the radioactive content of rocks from the strength and nature of airborne anomalies and great progress has been made in this direction. Moxham [214] states that the content of sources with large dimensions, e.g. marine phosphate deposits, can be determined within a few thousandths of a per cent uranium (or its equivalent). The accuracy decreases with the areal extent of the source.

It should be realized that on account of the shallow depths (generally 1 m or less) sensed by radioactivity, the method is often better described as a technique for mapping soils, except where barren rock is exposed. However, Schwarzer and Adams have reported in an extremely interesting paper [219] that the concentrations of K, U and Th determined from the air in their survey in Payne County, Oklahoma, USA, suggested that the 'signatures' of these elements in the rock are preserved in the overlying *in situ* soils. In such a case it should be possible to map lithologies and lithologic contacts by airborne radioactivity measurements.

9 Miscellaneous methods and topics

As the requirements of modern oil and mining exploration, hydrological investigations and civil engineering have grown, a number of special techniques suited to particular problems have been proposed and applied with varying degrees of success. The purpose of this chapter is to mention a few of these techniques and also discuss some topics which are of common interest in all geophysical surveys.

9.1 Borehole magnetometer

The measurement of electric resistivity, self-potential and elastic wave velocities in boreholes has already been mentioned (pp. 128, 219). It may be added that magnetic intensities too can be measured in a borehole by lowering a flux-gate or a proton magnetometer in it.

The borehole magnetometer has been used chiefly in iron-ore prospecting as an auxiliary instrument. If, for instance, a borehole has failed to encounter an expected ore zone, magnetic measurements can frequently reveal whether the zone is present in the vicinity of the hole and also indicate its distance. Weak and erratic mineralization often makes it difficult or impossible to establish the precise limits of a magnetic impregnation zone by an inspection of the drill cores alone. In such cases, the borehole magnetometer may be of considerable assistance.

The instrument can also be employed sometimes for correlating zones of the same grade in different parts of a magnetite ore deposit.

9.2 Gamma-ray logging

This technique utilizes the natural radioactivity of rocks and is used chiefly for correlating sedimentary strata in petroleum

233

prospecting. The apparatus consists, in principle, of a gamma-ray detector and its preamplifier suspended in a borehole by means of a waterproof electrical cable. The output of the detector is further amplified on the surface and recorded continuously as the detector is lowered (or raised) in the hole.

Generally speaking, shales and shaly sandstones show a very high radioactivity (the highest in sedimentary rocks) while salt, coal, anhydrite, limestones, quartz sands, etc. are weakly radioactive. Thus in a gamma-ray log the peaks in the intensity will correspond, in general, to shales while the lows will indicate the presence of limestones, salt, etc.

The gamma-ray log has also been used in uranium prospecting [220].

9.3 Neutron logging

There are two borehole methods known as neutron–gamma and neutron–neutron logging which employ neutrons. A convenient source of neutrons often used is a mixture of radium and powdered beryllium. The beryllium is bombarded by α-particles from the radium and fast neutrons are produced according to the reaction

$$_4^9\text{Be} + {}_2^4\text{He} \rightarrow {}_6^{12}\text{C} + {}_0^1\text{n} + \text{energy}$$

In collisions with nuclei neutrons are gradually slowed down until they reach thermal velocities. Hydrogen nuclei are particularly efficient in producing such 'thermal' neutrons and then capturing them according to the reaction

$$_0^1\text{n} + {}_1^1\text{H} \rightarrow {}_1^2\text{H} + \gamma$$

with the production of γ-radiation.

In rocks, hydrogen nuclei are present in oil, water, natural gas, etc. and the γ-radiation to which they give rise can be detected by lowering a neutron source just ahead of a gamma-ray counter.

In the neutron–neutron method the intensity of the neutrons scattered by the hydrogen nuclei, rather than the intensity of the gamma radiation due to their capture, is detected. Neutrons do not produce appreciable ionization so that a special device is needed for their detection. One such device is a Geiger tube filled with boron trifluoride gas. The neutrons react with the boron transforming it into lithium and releasing an α-particle which in turn ionizes the gas and reveals the presence of neutrons.

In passing through matter with a high hydrogen content the

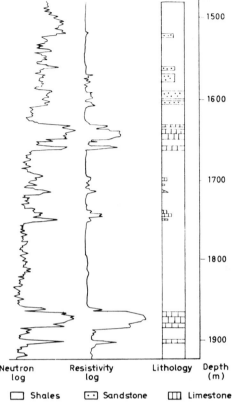

| Neutron log | Resistivity log | Lithology | Depth (m) |

☐ Shales   ⊡ Sandstone   ⊞ Limestone

Fig. 95. Neutron and resistivity logging.

neutrons are slowed down and captured by the hydrogen nuclei at a very small distance from the source. On the other hand, if the hydrogen content is low they travel a relatively large distance before reaching thermal velocities.

The number of neutrons arriving at the detector will be less when hydrogen is present than when it is absent. A decrease in the neutron response therefore indicates the presence of hydrogen in the rock and it may be assumed that this in turn is due to oil, water or gas, all of which have an abundance of hydrogen nuclei.

A neutron log and, for comparison, the resistivity log in the same sedimentary column are shown in Fig. 95.

9.4 Geothermal methods

Owing to the radioactivity of the crustal rocks heat is being continuously transported to the surface of the earth from its interior at a mean rate of about 50 mW/m^2. The lateral variations of this heat flow over the surface of the earth are small. If there are any appreciable local variations they are superimposed on the thermal effects due to vegetation, microclimates, etc. The latter are so large that surface temperature measurements cannot (as a rule) be used for deducing the thermal conductivity of rocks at depth or for determining the position of structures such as buried domes, anticlines, etc. They have, however, been successfully applied in finding fissures and cracks along which convective transfer of heat has taken place from the depth through the agency of water [221].

Also, Pooley and van Steveninck have recently shown in an excellent paper [222] that temperatures only about 2 m below the ground surface are substantially free from the above effects and measurements in short drillholes are capable of revealing shallow salt domes and similar structures by anomalies of the order of 1–2° C (50–100 times the measurement accuracy!).

Contrary to the behaviour of the total heat flow, the vertical gradient of the temperature in the earth varies within wide limits (5–70° C/km) depending upon the thermal conductivity of rock formations and temperature logs in deep boreholes can be used with advantage in correlating stratigraphic horizons.

9.5 Geochemical prospecting

Indications of oil, ore and other minerals can be sought by chemical analyses of the solid, liquid, and gaseous substances of which the earth's crust is composed. Such geochemical prospecting can be carried out in the bedrock, in loose overburden or in the uppermost layer of the earth's crust, namely the surface soils [223, 224].

The analyses of the gases slowly diffusing from very great depths to the surface of the earth has been employed in petroleum prospecting for quite some time. In this case the contents of hydrocarbons like methane, ethane, propane, etc. (which are intimately associated with oil) are determined in samples of gas collected just below the surface of the earth.

In mineral prospecting the object is to determine the metallic contents at different places on the earth. Metal ions can migrate through a number of agencies like weathering, leaching, wind transport, etc. several tens or even hundreds of metres from the parent ore body and create 'haloes' of mineralization around it. The metal concentrations in such haloes are in general very small and refined colorimetric and spectrographic methods are needed for reliable determinations.

The tracing of haloes is reported to have led to the discovery of a large number of mineralized zones in different parts of the world.

The phenomenon of the migration of metals within the earth's crust is an exceedingly complex one. Nevertheless, great progress has been made in its study, particularly in the USSR [223]. The importance of geochemical prospecting will be apparent from the fact that many metals like molybdenum, tungsten, vanadium, etc. which are of vital importance in modern industry, occur in the earth's crust in quantities far too small to appreciably alter the physical properties of the host rock so that their detection by geophysical means is out of the question.

9.6 Optimum point and line spacing

It is clear that the amount of information that can be extracted about sub-surface features from geophysical measurements will be greater the denser the network of the observation points. A maximum amount of information will be available when the points are infinitely close to each other. However, this ideal cannot be realized, firstly due to the finite size of any instrument and secondly due to economic and practical considerations. Therefore a balance must be struck between the extent of information desired and the amount of detail to be mapped.

Geophysical observations are most conveniently made at a number of consecutive points along a set of parallel straight lines. It is advantageous to stake the lines perpendicular to the known or presumed geological strike, which is the direction of the trace made on the earth's surface by inclined strata.

The geological conditions tend to be uniform over relatively large distances along a line in the strike direction but may vary considerably within short distances in the perpendicular direction (Fig. 96). If the lines of measurement are normal to the strike their mutual distance may be kept large while the geophysical observations are taken along each of them at relatively close intervals.

The distance between observation points must be adjusted to the anticipated depth of the feature one is seeking. The anomalies of shallow features are narrow while those of deep-lying features extend over greater distances. As a rule geophysical observations cannot be expected to yield information about features whose

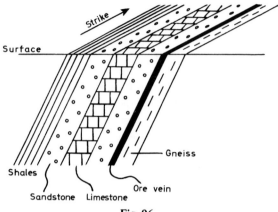

Fig. 96.

depth is much smaller than the distance between the observation points.

The optimum distance between the lines is dependent upon the length of the sub-surface feature, or more correctly, upon the estimated length of its measurable surface anomaly. If the distance is too large there may be no line which crosses a short anomaly and the corresponding feature may be undetected. On the other hand, if the line distance is too small there may be quite a number of lines crossing a long anomaly. In general, such lines tend to add to the cost of the survey without providing very much additional information.

Mathematically, the problem of optimum line spacing is that of finding the probability of detecting a geophysical anomaly of given dimensions by surveys with a particular line spacing. This is actually an old problem in 'continuous geometric probabilities' [225] whose solution is well known. For example, if the length of an unknown anomaly is L, the probability that one of a set of lines with a spacing $S(\geqslant L)$ crosses the anomaly is $2L/\pi S$. If a number of ore bodies each of which produces an anomaly of length L are randomly distributed in an area the probability of detecting them by a survey with a line spacing $S = L$ is 0.636.

For $S \leqslant L$ the probability that at least one line crosses the anomaly is

$$\frac{2L}{\pi S} [1 - \sqrt{(1 - S^2/L^2)}] + \frac{2}{\pi} \cos^{-1}(S/L)$$

If the length of the ore bodies is, say, 100 m the probability of detecting them all with a line spacing is 20 m is 0.974.

Other aspects of optimum line spacing have been considered by Agocs [226]

9.7 Position location in airborne surveying

The interpretation of airborne geophysical anomalies and the subsequent ground follow-up require a knowledge of the altitude of the aircraft and its position in the horizontal plane.

The altitude is measured by a barometric or a radio-altimeter and is recorded continuously alongside the data representing the results of the geophysical measurements. The position of the aircraft in the horizontal plane is much more difficult to determine. If detailed large-scale maps or good aerial photographs are available the following method is commonly used for this purpose.

The line to be flown is drawn on the map or the photograph and the pilot takes his bearing on a suitable distant landmark on the line. A navigator sitting beside the pilot watches for details on the flight lines such as rivers, lakes, brooks, railways, houses, etc. At intervals, when flying over a clearly identifiable feature he makes a mark on the map and at the same time presses a signal button which puts another mark on a trace on the recording paper. The succession of marks on the map and the recording paper can be correlated at the end of the flight. The navigator also informs the pilot of any adjustments necessary in the course of the plane.

If maps or air photographs are not available a camera is installed in the plane and exposures are made at regular intervals during flight. The number and the instant of an exposure are, of course, automatically registered on the recording paper. The interval between exposures is usually such that a certain amount of overlap between two adjacent photographs is obtained.

The above position location systems function satisfactorily only so long as there are a sufficient number of distinct topographic features in the area. In flying over dense jungles, deserts, seas, etc. the paucity or absence of such features severely limits the value of photographic methods and some form of radio location must be used. Some of these methods employ radar pulses while others are operated on continuous radio frequency waves. The general principle of the latter methods (Decca, Raydist, Lorac, etc.) is to send out radio waves simultaneously from two transmitter stations at fixed and precisely known locations and measure the phase difference between them at the receiver station (aeroplane). The phase difference is essentially a measure of the difference between the travel times of the two electromagnetic waves and is therefore proportional to the difference of the transmitter distances from the aeroplane.

Another method of radio location uses the Doppler effect. Four radio beams directed towards the ground are sent out making a small angle with the vertical from a transmitter in the aeroplane, two in the forward and aft directions and two sideways. The Doppler frequency shifts of the reflected waves due to the motion of the aeroplane are made to yield the deviations of the plane from the flight line. Thus, starting from a given station and a given initial bearing the actual flight path can be reconstructed from the continuous record of the Doppler shifts.

9.8 Composite surveys

The choice of the method for a geophysical survey is guided by a number of considerations such as the object of the survey, the geology and topography of the area to be investigated and the type of information sought about the sub-surface. The last mentioned factor is, of course, of fundamental importance.

It is frequently advantageous to use two or more suitable methods within a given area. Such a composite survey gives additional information about the physical properties of the sub-surface and helps (in combination with the geological knowledge) to reduce the uncertainty inherent in the interpretation of geophysical data.

For example, an electromagnetic anomaly may indicate an electric conductor but sometimes the conductor may be so good that the strong eddy currents in it will screen other less good conductors in the vicinity from the primary field and these will not be detected. In such a case an earth resistivity (or a self-potential) survey will often reveal their presence. A magnetic survey will further eliminate (or confirm) the possibility whether these conductors contain magnetite (pyrrhotite) or are composed purely of non-magnetic conducting minerals. A gravity survey will often help to distinguish between compact, massive ore bodies and zones of poor, disseminated mineralization since the density contrast with the host rock will be large in the first case and small in the second.

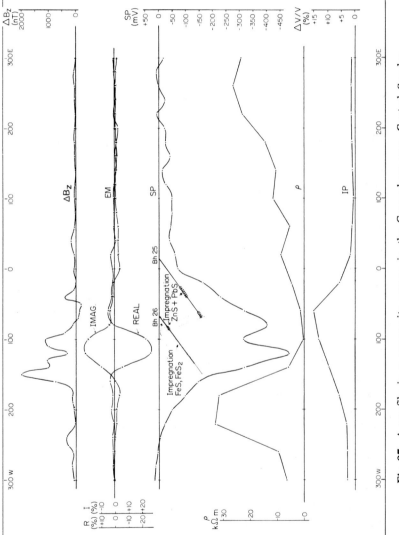

Fig. 97. A profile in a composite survey in the Garpenberg area, Central Sweden.

Faults, domes, anticlines and other geological structures indicated by a seismic survey can also be 'screened' by gravity observations to obtain information about the density contrasts involved, while magnetic measurements can frequently reveal diabase and gabbro dikes which may not be indicated by the seismic survey owing to their insufficient velocity contrast with the country rock.

Examples of the advantages of composite surveys can be easily multiplied. The results along one of the profiles in a composite electromagnetic, magnetic, self-potential, earth resistivity and

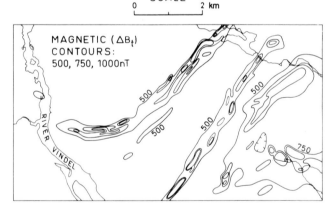

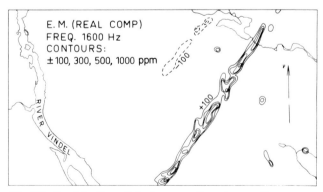

Fig. 98. Combination of electromagnetic and magnetic surveys.

induced polarization survey in a lead–zinc–copper sulphide mineralization district in central Sweden are shown in Fig. 97, where the scale is in metres. Many of the above-mentioned effects will be recognized in this example. For instance, the electromagnetic indication is due to a thin sulphide vein but the resistivity survey reveals a broad low-resistivity zone of which only a part seems, to judge from the IP profile, to be electronically conducting. The SP profile differentiates two tops, corresponding probably to two separate sulphide concentrations within the broad impregnation zone.

Another example in which a combination of two methods supplies extra information is shown in Fig. 98. The upper part shows aeromagnetic anomalies. We notice two long and parallel, narrow belts to the west. On the airborne electromagnetic map (lower part) only one of them appears strongly. Evidently, the north-western magnetic anomaly must be caused by a long ultrabasic dike whereas the other anomaly is due to a pyrrhotite-bearing black-shale horizon which, incidently, is a key horizon for ore prospecting in this region. It should also be noticed that a small part of the ultrabasic dike does appear on the electromagnetic map giving weak negative anomalies.

With the coil-configuration used (wing-tip mounted coils with dipole axes in the flight direction) tabular conductors give positive anomalies as, for instance, the shale horizon does. The weak negative electromagnetic anomalies of the ultrabasic dike are actually due to the magnetic permeability of the dike (cf. Section 6.15). The eastern magnetic complex, on the other hand, does not give electromagnetic anomalies, neither positive nor negative. It is clear that not only is this complex a poor electric conductor but also that it has a low magnetic permeability. Presumably this is also basic rock but its content of magnetic material is much smaller than that of the ultrabasic dike.

Several examples of composite surveys and references to case histories will be found elsewhere [66, 147, 186].

APPENDIX 1

Magnetic potential

According to Green's theorem, if U, V are two functions with sufficient differentiability in a region R (volume element dv), then

$$\iiint_R (U\nabla^2 V - V\nabla^2 U)\, dv = \iint_S \left(U\frac{\partial V}{\partial n} - V\frac{\partial U}{\partial n} \right) dS \quad \text{(A 1.1)}$$

where S is the surface bounding R and $\partial/\partial n$ denotes differentiation with respect to the outward normal to dS.

Let R denote the *whole* of space, M_x, M_y, M_z the components of magnetization at any point of it, and ϕ the magnetic potential. Put $U = \phi$, $V = 1/r$ where r is the distance between a definite point $Q(\xi, \eta, \zeta)$ in R and an arbitrary point $P(x, y, z)$ in R.

Surround Q by a sphere Σ with Q as centre and let R_1 denote the region of R external to Σ. The surface σ of Σ is one boundary of R_1, the other boundary being, of course, at infinity. It can be verified by direct differentiation that $\nabla^2 V = 0$ in R_1. Then from (A 1.1),

$$\iiint_{R_1} -\frac{\nabla^2 \phi}{r}\, dv = \iint_\sigma \left(\phi\frac{\partial}{\partial n}\frac{1}{r} - \frac{1}{r}\frac{\partial \phi}{\partial n} \right) d\sigma$$

Since the direction of n is opposite to the radius vector ϵ from Q to $d\sigma$, $\partial/\partial n = -\partial/\partial \epsilon$. Furthermore $r = \epsilon$ on σ, so that

$$\iiint_{R_1} -\frac{\nabla^2 \phi}{r}\, dv = \iint_\sigma \left(\frac{\phi}{\epsilon^2} + \frac{1}{\epsilon}\frac{\partial \phi}{\partial \epsilon} \right) d\sigma$$

But $d\sigma = \epsilon^2\, d\Omega$ where $d\Omega$ is the solid angle subtended by $d\sigma$ at Q. Hence

$$\iiint_{R_1} -\frac{\nabla^2 \phi}{r}\, dv = \iint_\sigma \left(\phi + \epsilon\frac{\partial \phi}{\partial \epsilon} \right) d\Omega$$

which reduces as $\epsilon \to 0$ (and consequently $R_1 \to R$) to

$$\iiint_R - \frac{\nabla^2 \phi}{r} \, dv = \phi(Q) \iint d\Omega = 4\pi\phi(Q) \qquad \text{(A 1.2)}$$

Since $\nabla^2 \phi = \text{div } \mathbf{M}$ (Equation 2.5) we get

$$\phi = -\frac{1}{4\pi} \iiint_R \frac{\text{div } \mathbf{M}}{r} \, dv \qquad \text{(A 1.3)}$$

where it is *implicit that* M *is continuous throughout* R.

It is more or less obvious that

$$\frac{\text{div } \mathbf{M}}{r} \equiv \frac{1}{r} \left(\frac{\partial M_x}{\partial x} + \frac{\partial M_y}{\partial y} + \frac{\partial M_z}{\partial z} \right)$$

$$= \text{div} \left(\frac{\mathbf{M}}{r} \right) - \Sigma M_x \frac{\partial}{\partial x} \left(\frac{1}{r} \right)$$

Hence

$$\phi = -\frac{1}{4\pi} \left[\int_R \text{div} \left(\frac{\mathbf{M}}{r} \right) \, dv - \int_R \Sigma M_x \frac{\partial}{\partial x} \left(\frac{1}{r} \right) dv \right]$$

If M is discontinuous on some surface bounding a volume R_1, encapsule the surface by two surfaces S_1, S_2 (dashed lines in Fig. 99) on either side of it, enclosing an infinitesimal volume $\Delta\tau$. Then

$$\phi = -\frac{1}{4\pi} \left[\int_{R_1} \frac{\text{div } \mathbf{M}}{r} \, dv + \int_{\Delta\tau} \frac{\text{div } \mathbf{M}}{r} \, dv + \int_{R_2} \frac{\text{div } \mathbf{M}}{r} \, dv \right]$$

$$= -\frac{1}{4\pi} \left[\int_{R_1} \text{div} \left(\frac{\mathbf{M}}{r} \right) \, dv + \int_{\Delta\tau} \text{div} \left(\frac{\mathbf{M}}{r} \right) \, dv + \int_{R_2} \text{div} \left(\frac{\mathbf{M}}{r} \right) dv \right.$$

$$- \int_{R_1} \Sigma M_x \frac{\partial}{\partial x} \left(\frac{1}{r} \right) \, dv - \int_{\Delta\tau} \Sigma M_x \frac{\partial}{\partial x} \left(\frac{1}{r} \right) \, dv$$

$$\left. - \int_{R_2} \Sigma M_x \frac{\partial}{\partial x} \left(\frac{1}{r} \right) \, dv \right] \qquad \text{(A 1.4)}$$

However, by the divergence theorem

$$\int_V \text{div} \left(\frac{\mathbf{M}}{r} \right) \, dv = \int_S \frac{lM_x + mM_y + nM_z}{r} \, dS \qquad \text{(A 1.5)}$$

where S is the bounding surface of a volume V and l, m, n are the direction cosines of the outward normal of dS. It is easy to see that S for R_1 is S_1; for $\Delta\tau$ it is S_1 and S_2 and for R_2 it is S_2.

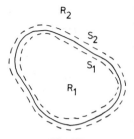

Fig. 99.

However dS_1, considered as an element of the bounding surface of R_1, has an outward normal that is exactly opposite to the outward normal when dS_1 is considered as an element bounding $\Delta\tau$. Hence the integrals over S_1 cancel each other. Similarly the integrals over S_2 cancel each other. Hence in view of (A 1.5) the first three terms on the right-hand side of (A 1.4) sum to zero, and when $\Delta\tau \to 0$ the equation reduces to

$$\phi = \frac{1}{4\pi} \iiint_R \left[M_x \frac{\partial}{\partial x} \left(\frac{1}{r} \right) + M_y \frac{\partial}{\partial y} \left(\frac{1}{r} \right) + M_z \frac{\partial}{\partial z} \left(\frac{1}{r} \right) \right] \, dv$$

(A 1.6)

It should be noted that the integral is to be taken over *all* space and not merely throughout the bounded volume R_1. Only if the subspace outside R_1 is non-magnetic ($M = 0$ outside R_1) does the integral reduce to one over R_1 alone.

APPENDIX 2

Transition energy in the alkali vapour magnetometer

By the quantum theory the projection of the angular momentum J_a of an alkali atom on a (weak) magnetic field must be a half-integral multiple of $h/2\pi$. Thus the possible orientations of J_a with respect to the field are given by

$$\cos \theta = \frac{mh}{2\pi J_a}$$

where m is half-integral.

Also, in the presence of a weak magnetic field the effective atomic magnetic moment m_a is in the direction of J_a.

The energy of a magnetic dipole that makes an angle θ with a magnetic field B is $m_a B \cos \theta$ or, for the alkali atom, $m_a B(mh/2\pi J_a)$. If a transition takes place from a state with $m = \frac{1}{2}$ to one with $m = -\frac{1}{2}$ the energy change involved is

$$\frac{m_a B h}{2\pi J_a}$$

and setting this equal to $h\nu$ we get

$$\nu = \frac{m_a B}{2\pi J_a}$$

But from the theory of the gyroscope this is also the frequency of precession of the atomic spin J_a around B. (It must be remarked that the above demonstration is incomplete for a variety of reasons, chiefly because it does not take account of the so-called Landé factor.)

APPENDIX 3

Magnetized sphere and a magnetic dipole

A3.1 Isotropic, homogeneous sphere subject to a uniform external magnetizing force H_0 (A m^{-1})

Since the potential due to magnetization is given by Equation (A 1.6) the net magnetizing force at any point $Q(\xi, \eta, \zeta)$ in, say, the x direction is

$$H_x = H_{0x} + \left(-\frac{\partial \phi}{\partial \xi}\right) \tag{A 3.1}$$

We shall calculate the components H_{ix}, H_{iy}, H_{iz} of the magnetizing force H_i when Q is inside the sphere, assuming a susceptibility κ for the sphere and zero for the outside medium so that $\mathbf{M} = \kappa \mathbf{H}_i$. Without loss of generality it may be assumed that x, y, z and ξ, η, ζ are rectangular coordinates with O, the centre of the sphere, as origin. Then from Equations (A 3.1) and (A 1.6),

$$H_{ix} = H_{ox} - \frac{\kappa}{4\pi} \frac{\partial}{\partial \xi} \iiint_{\text{sphere}} \left(H_{ix} \frac{\partial}{\partial x}\left(\frac{1}{r}\right)\right.$$

$$\left. + H_{iy} \frac{\partial}{\partial y}\left(\frac{1}{r}\right) + H_{iz} \frac{\partial}{\partial z}\left(\frac{1}{r}\right)\right) \, \mathrm{d}v \tag{A 3.2}$$

with two similar equations for H_{iy}, H_{iz}.

To solve Equation (A 3.2) for H_{ix}, assume as a first guess that H_i is constant within the sphere. Then the three integrals to be evaluated on the right-hand side become

$$\iiint_{\text{sphere}} \frac{\partial}{\partial x}\left(\frac{1}{r}\right) \mathrm{d}v, \quad \iiint_{\text{sphere}} \frac{\partial}{\partial y}\left(\frac{1}{r}\right) \mathrm{d}v, \quad \iiint_{\text{sphere}} \frac{\partial}{\partial z}\left(\frac{1}{r}\right) \mathrm{d}v$$

These can be converted into the surface integrals

$$\iint \frac{l\,dS}{r}, \quad \iint \frac{m\,dS}{r}, \quad \iint \frac{n\,dS}{r}$$

where l, m, n are the direction cosines of the surface element dS.

The integrals are elementary and each can be readily evaluated by choosing the lines $O\xi$, $O\eta$ or $O\zeta$ as the respective polar axis of a spherical polar coordinate system. Since

$$\frac{1}{r} = \frac{1}{\{(x-\xi)^2 + (y-\eta)^2 + (z-\zeta)^2\}^{1/2}}$$

the integrals turn out to be $4\pi\xi/3$, $4\pi\eta/3$ and $4\pi\zeta/3$ (when Q is inside the sphere) so that from (A 3.2)

$$H_{ix} = H_{ox} - \frac{1}{3}\kappa H_{ix}$$

or

$$H_{ix} = \frac{H_{ox}}{1 + \kappa/3}$$

By symmetry

$$H_{iy} = \frac{H_{oy}}{1 + \kappa/3}, \quad H_{iz} = \frac{H_{oz}}{1 + \kappa/3}$$

It will be obvious on a moment's reflection that if these values of H_{ix}, H_{iy}, H_{iz} are used as the next guess in Equation (A 3.2) the new values will again be the same. Hence it follows that the internal field in the sphere is uniform, as also is its magnetization intensity

$$\mathbf{M} = \frac{\kappa \mathbf{H}_0}{1 + \kappa/3} \tag{A 3.3}$$

The factor $1/3$ in the denominator is *the demagnetization factor*.

A3.2 Magnetic potential of a uniformly magnetized sphere at external points

There is no loss of generality in rotating the coordinate axes in Equation (A 1.6) so that the x-axis, say, coincides with the direction of $\mathbf{M}$. Then $M_y = M_z = 0$ and $M_x = M$. Hence

$$\phi = \frac{M}{4\pi} \iiint \frac{\partial}{\partial x}\left(\frac{1}{r}\right) dv$$

The quickest way to evaluate the integral here is to note that formally it represents the x-component of the external gravitational attraction of a sphere of unit density. Its value is therefore $\frac{4}{3}\pi b^3 (\cos\theta / R^2)$ where b is the radius of the sphere, R is the distance of Q from O, the centre of the sphere, and θ is the angle between OQ and the x-axis, that is, M. Thus

$$\phi = \frac{m}{4\pi} \frac{1}{R^2} \cos\theta$$

where $m = \frac{4}{3}\pi b^3 M$ is the magnetic moment of the sphere, a result that, after comparison with Equation (2.8), shows that the potential of a homogeneously magnetized sphere at external points is the same as that of a dipole of identical moment placed at the sphere's centre.

For a sphere in a uniform field, then,

$$m = \frac{4\pi}{3} b^3 \frac{3\kappa H_0}{3 + \kappa} \quad (A\ m^2)$$

A3.3 Magnetizing forces parallel and perpendicular to a dipole

Starting from Equation (2.8) we have

$$-\frac{\partial\phi}{\partial r} = \frac{m}{4\pi} \frac{2}{r^3} \cos\theta$$

and

$$-\frac{1}{r}\frac{\partial\phi}{\partial\theta} = \frac{m}{4\pi} \frac{1}{r^3} \sin\theta$$

for the magnetizing forces in the direction of and perpendicular to the radius vector r from the dipole. Each of these can be resolved in the direction parallel and perpendicular to the dipole and it is quite easily shown that the magnetizing forces parallel and perpendicular to the dipole are

$$H_{\parallel} = \frac{m}{4\pi} \frac{1}{r^3} (3 \cos^2\theta - 1) \tag{A 3.4a}$$

$$H_{\perp} = \frac{m}{4\pi} \frac{3 \sin\theta \cos\theta}{r^3} \tag{A 3.4b}$$

APPENDIX 4

Magnetic potential of a linear dipole

Let σ be the cross-section of a linear dipole extending, say, in the y-direction from $y = y_1$ to $y = y_2$. Then dv in Equation (A 1.6) $= \sigma \, dy$. If the point of observation Q is chosen as the origin of coordinates and if M_x, M_y, M_z are independent of y, Equation (A 1.6) can be written immediately as

$$\phi = \frac{1}{4\pi} M_x \sigma \frac{\partial}{\partial x} \int_{y_1}^{y_2} \frac{dy}{(x^2 + y^2 + z^2)^{1/2}} + \frac{1}{4\pi} M_y \sigma \int_{y_1}^{y_2} \partial\left(\frac{1}{r}\right)$$

$$+ \frac{1}{4\pi} M_z \sigma \frac{\partial}{\partial z} \int_{y_1}^{y_2} \frac{dy}{(x^2 + y^2 + z^2)^{1/2}}$$

$$= \frac{1}{4\pi} M_x \sigma \frac{\partial}{\partial x} \ln\left(\frac{R_2 + Y_2}{R_1 - Y_1}\right) + \frac{1}{4\pi} M_y \sigma \left(\frac{1}{R_2} - \frac{1}{R_1}\right)$$

$$+ \frac{1}{4\pi} M_z \sigma \frac{\partial}{\partial z} \ln\left(\frac{R_2 + Y_2}{R_1 - Y_1}\right) \qquad (A\ 4.1)$$

where $R_2 = (x^2 + y_2^2 + z^2)^{1/2}$ and $R_1 = (x^2 + y_1^2 + z^2)^{1/2}$. ($R_1$, R_2 are the distances of Q from the end-points of the dipole).

For an infinitely long dipole ($y_2 = \infty$, $y_1 = -\infty$) the second term vanishes whereas the limit of the logarithmic terms can be shown to be $-2 \ln(x^2 + z^2)^{1/2} + C_\infty$, where C_∞ is an infinite constant, after expressing R_1, R_2 as

$$y_1\left(1 + \frac{x^2 + z^2}{y_1^2}\right)^{1/2} \quad \text{and} \quad y_2\left(1 + \frac{x^2 + z^2}{y_2^2}\right)^{1/2}$$

Hence, for an infinitely long dipole striking in the y-direction

$$\phi = -\frac{1}{4\pi} 2M_x \sigma \frac{\partial}{\partial x} \ln\rho - \frac{1}{4\pi} 2M_z \sigma \frac{\partial}{\partial z} \ln\rho + C_\infty \qquad (A\ 4.2)$$

251

where $\rho = (x^2 + z^2)^{1/2}$ is obviously the shortest distance from Q to the dipole.

$M_x\sigma$, $M_z\sigma$ are the magnetic moments μ_x, μ_z in the x- and z-directions per unit length of the dipole.

APPENDIX 5

Magnetic anomaly of a thick sheet

A5.1 Derivation of Equation 2.37

The infinitely long thick sheet in Fig. 100 may be considered to be built up of elementary thin sheets of thickness $d\epsilon$ as shown. If $d(\Delta B_h)$, $d(\Delta B_z)$ are the horizontal and vertical flux densities of an elementary sheet, the flux density in the direction $(l, 0, n)$ is given by

$$d\{\Delta B(l, 0, n)\} = l\,d(\Delta B_h) + n\,d(\Delta B_z)$$

If ξ is the horizontal coordinate of the elementary sheet we get from Equations (2.34) and (2.35)

$$d(\Delta B) = \frac{\mu_0}{4\pi} 2d\epsilon \left[l \frac{(x - \xi)M'_\parallel + aM'_\perp}{a^2 + (x - \xi)^2} + n \frac{aM'_\parallel - (x - \xi)M'_\perp}{a^2 + (x - \xi)^2} \right]$$

However $d\epsilon = d\xi \sin\theta$ and the above equation may be integrated from $\xi = -b/2$ to $\xi = +b/2$. The integrations are elementary. Noting further (cf. Fig. 10) that $M'_\parallel = M' \cos(\theta - i')$ and $M'_\perp = M' \sin(\theta - i')$ the result stated in Equation (2.37) is obtained without difficulty.

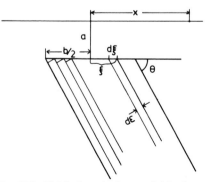

Fig. 100. Thick sheet built up of thin sheets.

APPENDIX 6

Potential of a point current electrode on the surface of a horizontally-layered earth

Let $n - 1$ layers rest on an nth 'layer', the infinite sub-stratum (Fig. 101). Choosing a cylindrical coordinate system R, θ, z with the electrode C as the origin and z positive downwards, Laplace's equation for the electric potential V in each layer can be written as

$$\frac{\partial^2 V}{\partial R^2} + \frac{1}{R}\frac{\partial V}{\partial R} + \frac{\partial^2 V}{\partial z^2} = 0 \qquad \text{(A 6.1)}$$

since, by symmetry, V is independent of θ.

Assuming $V(R, z) = F(R)G(z)$ where F is a function of R only and G of z only, the equation is separated into the two equations

$$\frac{d^2 G}{dz^2} - \lambda^2 G = 0 \qquad \text{(A 6.3a)}$$

$$\frac{d^2 F}{dR^2} + \frac{1}{R}\frac{dF}{dR} + \lambda^2 F = 0 \qquad \text{(A 6.3b)}$$

where λ is a constant independent of R and z.

Equation (A 6.3b) is Bessel's equation of order zero with fundamental solutions $J_0(\lambda R)$, $Y_0(\lambda R)$. $J_0(\lambda R)$ is finite for $R \to \infty$ only if λ is real, and since $Y_0(\lambda R)$ is always ∞ for $R \to \infty$, it must be rejected. The solutions of (A 6.3a) are $\exp(-\lambda z)$ and $\exp(\lambda z)$.

The most general solution for the potential in any layer j ($j \neq 1$ or n) is then

$$V_j = \int_0^\infty \{A_j(\lambda)e^{-\lambda z} + B_j(\lambda)e^{\lambda z}\} J_0(\lambda R)d\lambda \qquad \text{(A 6.4)}$$

On account of the occurrence of $\exp(\lambda z)$ as well as $\exp(-\lambda z)$, there is no loss of generality in restricting λ to positive real values only.

254

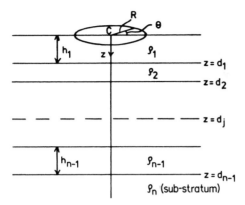

Fig. 101. Point electrode on a stratified earth and the cylindrical coordinate system.

For $j = n$, the term in $\exp(\lambda z)$ must be excluded since the potential in the sub-stratum must be finite as $z \to \infty$. Thus

$$V_n = \int_0^\infty A_n(\lambda)e^{-\lambda z} J_0(\lambda R)d\lambda \qquad \text{(A 6.5)}$$

In the topmost layer ($j = 1$), V_1 is the sum of the normal potential (cf p. 107)

$$V_0 = \frac{I\rho_1}{2\pi} \frac{1}{(R^2 + z^2)^{1/2}}$$

and a disturbance potential

$$V_d = \int_0^\infty A_1(\lambda)(e^{-\lambda z} + e^{\lambda z})J_0(\lambda R)d\lambda$$

The coefficients of $\exp(-\lambda z)$ and $\exp(\lambda z)$ in V_d are equal since $(1/\rho_1)(\partial V_d/\partial z) = 0$ at $z = 0$ (no current flow across the earth's surface except at C). Using Lipschitz's integral in Bessel function theory for V_0 we have

$$V_1 = \frac{I\rho_1}{2\pi} \int_0^\infty e^{-\lambda z} J_0(\lambda R)d\lambda + \int_0^\infty A_1(\lambda)(e^{-\lambda z} + e^{\lambda z})J_0(\lambda R)d\lambda \qquad \text{(A 6.6)}$$

From the continuity of V_j and the current density normal to the layer interfaces we have, for depths $z = d_j$ ($j = 1, 2, \ldots n - 1$),

$$V_{j-1} = V_j \qquad \text{(A 6.7a)}$$

$$\frac{1}{\rho_{j-1}} \frac{\partial V_{j-1}}{\partial z} = \frac{1}{\rho_j} \frac{\partial V_j}{\partial z} \qquad \text{(A 6.7b)}$$

There are altogether $2n - 2$ unknown functions $A(\lambda)$, $B(\lambda)$ to be determined ($2n - 4$ in (A 6.4) and 1 each in (A 6.5) and (A 6.6)). They can be determined by solving the system of $2n - 2$ linear equations obtained from the conditions (A 6.7). The solution is tedious but straightforward. For the details reference may be made to Kofoed's monograph [83]. Identifying $A_1(\lambda)$ as $(I\rho_1/\pi)K(\lambda)$ and R as r along the ground surface, it is easily seen that Equation (A 6.6) is the same as Equation (4.7).

$K(\lambda)$ is a function of the layer parameters, and recurrence formulae for building it for any number of layers on top of each other, starting from its expression for two layers, can be found by solving the system of Equations (A 6.7). Here the recurrence formulae for the transform $T(\lambda)$ of Equation (4.10) will be given instead, in terms of the *layer thicknesses* h_j ($j = 1, \ldots n - 1$), rather than the interface depths d_j.

For a layer (ρ_{n-1}, h_{n-1}) on top of a sub-stratum (ρ_n)

$$T_{n-1}(\lambda) = \rho_{n-1} \frac{1 - k_{n-1}u_{n-1}}{1 + k_{n-1}u_{n-1}}$$

where

$$u_{n-1} = \exp(-2h_{n-1}\lambda)$$
$$k_{n-1} = (\rho_{n-1} - \rho_n)/(\rho_{n-1} + \rho_n)$$

For the transform T_j for a layer (ρ_j, d_j) on top of the sequence ($\rho_{j+1}, \ldots \rho_n; h_{j+1} \ldots h_n$) with the transform T_{j+1}, we have

$$T_j(\lambda) = \frac{W_j(\lambda) + T_{j+1}(\lambda)}{1 + W_j(\lambda)T_{j+1}(\lambda)/\rho_j^2} \; ; j = n - 2, n - 3, \ldots 2, 1$$

where

$$W_j(\lambda) = \rho_j \frac{1 - u_j}{1 + u_j}$$

Starting from $T_{n-1}(\lambda)$ the transform $T_1(\lambda) = T(\lambda)$ of Equation (4.10) can be obtained by recursive application of the expression for $T_j(\lambda)$.

APPENDIX 7

Fourier transforms and convolution

A7.1 Fourier transforms

Any arbitrary function $f(t)$ (satisfying certain conditions) can be synthesized from a number of sine and cosine waves of different frequencies each with a characteristic amplitude and phase. If $f(t)$ is periodic with the period T the frequencies are discrete multiples of the fundamental frequency $1/T$. If $f(t)$ is non-periodic the frequencies are infinitesimally close to each other and range from 0 to ∞. For mathematical convenience, however, they are usually taken to range continuously from $-\infty$ to $+\infty$. The function $F(\nu)$ giving the amplitude of the wave of frequency ν is known as the amplitude spectrum. It can be obtained, as will appear below, from $f(t)$ while, conversely, if $F(\nu)$ is given $f(t)$ can be synthesized by essentially the same process.

We shall start with a periodic function $f(t)$ and express it by means of a Fourier series:

$$f(t) = \sum_{m=0}^{\infty} \left\{ a_m \, \mathrm{e}^{\mathrm{i}(2\pi m/T)t} + b_m \, \mathrm{e}^{-\mathrm{i}(2\pi m/T)t} \right\}$$

It is easily shown by multiplying both sides by $\exp(\mathrm{i}(2\pi n/T)t)$ and $\exp(-\mathrm{i}(2\pi n/T)t)$ and integrating from $-T/2$ to $+T/2$ that

$$a_m = (1/T) \int_{-T/2}^{T/2} f(t') \, \mathrm{e}^{-\mathrm{i}(2\pi m/T)t'} \, \mathrm{d}t'$$

$$b_m = (1/T) \int_{-T/2}^{T/2} f(t') \, \mathrm{e}^{\mathrm{i}(2\pi m/T)t'} \, \mathrm{d}t'$$

Hence

$$f(t) = (1/T) \sum_{m=0}^{\infty} \int_{-T/2}^{T/2} f(t') \left\{ \mathrm{e}^{\mathrm{i}(2\pi m/T)(t-t')} + \mathrm{e}^{-\mathrm{i}(2\pi m/T)(t-t')} \right\} \mathrm{d}t'$$

257

Since the bracketed expression on the right hand side is an even function of m it is obvious that we may also write

$$f(t) = (1/2T) \sum_{m=-\infty}^{\infty} \int_{-T/2}^{T/2} f(t') \left\{ e^{i(2\pi m/T)(t - t')} + \right.$$
$$\left. + e^{-i(2\pi m/T)(t - t')} \right\} dt'$$

Moreover, since $\exp(x) - \exp(-x)$ is an odd function of x we obviously have

$$0 = (1/2T) \sum_{m=-\infty}^{\infty} \int_{-T/2}^{T/2} f(t') \left\{ e^{i(2\pi m/T)(t - t')} \right.$$
$$\left. - e^{-i(2\pi m/T)(t - t')} \right\} dt'$$

Adding these two equations we get

$$f(t) = (1/T) \sum_{m=-\infty}^{\infty} e^{i(2\pi m/T)t} \int_{-T/2}^{T/2} f(t') e^{-i(2\pi m/T)t'} dt'$$

If $f(t)$ is non-periodic its period may be said to be infinite. Put $m/T = \nu$ and $1/T = d\nu$ and let $T \to \infty$. Then, from the definition of the Riemann integral, the sum may be replaced by an integral. Hence

$$f(t) = \int_{-\infty}^{\infty} d\nu \left[e^{i2\pi\nu t} \left\{ \int_{-\infty}^{\infty} f(t') e^{-i2\pi\nu t'} dt' \right\} \right]$$

Let

$$F(\nu) = \int_{-\infty}^{\infty} f(t) e^{-i2\pi\nu t} dt \qquad (A\ 7.1)$$

Then it follows that

$$f(t) = \int_{-\infty}^{\infty} F(\nu) e^{i2\pi\nu t} d\nu \qquad (A\ 7.2)$$

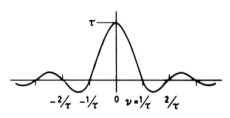

Fig. 102. Fourier spectrum of a square-wave pulse.

$F(\nu)$ is known as the Fourier transform of $f(t)$ while $f(t)$ is called the inverse Fourier transform of $F(\nu)$. If the one is given the other can be obtained from (A 7.1) or (A 7.2) respectively.

A square-wave pulse (p. 172) of duration τ and strength 1 is defined by the equations

$$f(t) = 1 \qquad |t| < \tau/2$$

$$= 1/2 \qquad |t| = \tau/2$$

$$= 0 \qquad |t| > \tau/2$$

It is easy to show by means of (A 7.1) that the corresponding amplitude spectrum (Fig. 102) is given by

$$F(\nu) = \tau \, \frac{\sin(\pi\nu\tau)}{(\pi\nu\tau)}$$

A7.2 Convolution

Let $f(t)$ represent the response of a linear system to a unit impulse, t seconds after the impulse has been applied. A continuous signal, whose strength at a time λ is $g(\lambda)$, may be considered to consist of an infinite number of successive very closely spaced impulses, each of strength $g(\lambda)d\lambda$. At a time t, the system's response to the impulse $g(\lambda)d\lambda$ is evidently $g(\lambda)f(t - \lambda)d\lambda$ since $t - \lambda$ is the time elapsed after the impulse $g(\lambda)d\lambda$. The total response of the system at time t is then

$$r(t) = \int_{-\infty}^{\infty} g(\lambda)f(t - \lambda)\,d\lambda \qquad \text{(A 7.3)}$$

$$= \int_{-\infty}^{\infty} g(t - \lambda)f(\lambda)\,d\lambda \qquad \text{(A 7.4)}$$

The integral in (A 7.3) or (A 7.4) is known as the convolution integral. The limits $-\infty$ and $+\infty$ are chosen merely for the sake of convenience in analysis. Physically, if $g(\lambda) \equiv 0$ for $\lambda < t_0$, say, and $f(t - \lambda) \equiv 0$ for $t - \lambda < t_1$, that is $\lambda > t - t_1$, the effective limits of integration will be simply t_0 and $t - t_1$.

Taking the Fourier transform of both sides of (A 7.4),

$$R(\nu) = \int_{-\infty}^{\infty} \int_{-\infty}^{\infty} g(t - \lambda) f(\lambda) e^{-i2\pi\nu t} \, dt \, d\lambda$$

$$= \int_{-\infty}^{\infty} \left[\int_{-\infty}^{\infty} g(t - \lambda) e^{-i2\pi\nu(t-\lambda)} \, dt \right] f(\lambda) e^{-i2\pi\nu\lambda} \, d\lambda$$

$$= \int_{-\infty}^{\infty} [G(\nu)] f(\lambda) e^{-i2\pi\nu\lambda} \, d\lambda$$

$$= G(\nu) \int_{-\infty}^{\infty} f(\lambda) e^{-i2\pi\nu\lambda} \, d\lambda$$

$$= G(\nu) F(\nu) \qquad\qquad\qquad\qquad \text{(A 7.5)}$$

Equation (A 7.5) shows that the Fourier transform of the convolution of two functions is equal to the product of the Fourier transforms of the functions. The same result follows if we start from (A 7.3).

A7.2.1 Analytic downward continuation (p. 82)

The formal solution of (3.16a) (p. 83) is easily obtained by using the above convolution theorem. Since

$$\frac{1}{r^3} = \frac{1}{\{(x - x_0)^2 + (y - y_0)^2 + z^2\}^{3/2}}$$

the right-hand side of (3.16a) may be regarded as a convolution (in two dimensions) of $1/r^3$ and $(z/2\pi) g(x_0, y_0, z)$. Instead of one frequency as in (A 7.1) we then have two spatial frequencies ν, μ (that is, wavenumbers = reciprocals of wavelengths) in the x- and y-directions. Taking the two-dimensional Fourier transform of both sides in (3.16a) we get, by the convolution theorem,

$$G(\nu, \mu) = G_0(\nu, \mu) e^{-\sqrt{(\nu^2 + \mu^2)} z} \qquad\qquad \text{(A 7.6)}$$

since it can be shown that the Fourier transform of $1/r^3$ is

$$\frac{2\pi}{z} \cdot e^{-\sqrt{(\nu^2 + \mu^2)} z} \cdot e^{-i2\pi(\nu x_0 + \mu y_0)}.$$

Therefore

$$G_0(\nu, \mu) = e^{\sqrt{(\nu^2 + \mu^2)} z} G(\nu, \mu)$$

Now, $g(x, y, 0)$ being the given function, $G(v, \mu)$ can be obtained by the numerical evaluation of two double integrals, namely,

$$\int_{-\infty}^{\infty} \int_{-\infty}^{\infty} g(x, y, 0) \, \frac{\cos}{\sin} \, (vx + \mu y) \, dx \, dy$$

for various values of v and μ. Then $g(x_0, y_0, z)$ is obtained as the inverse Fourier transform of $G_0(v, \mu)$. Towards this end we must numerically evaluate the two double integrals

$$\int_{-\infty}^{\infty} \int_{-\infty}^{\infty} e^{\sqrt{(v^2 + \mu^2)} z} \, G(v, \mu) \, \frac{\cos}{\sin} \, (vx_0 + \mu y_0) \, dv \, d\mu$$

for various values of x_0, y_0.

It should be noticed here that the amplitudes of all frequencies in $g(x, y, 0)$ are magnified in downward continuation on account of the exponential factor. The high frequencies (rapid variations) are magnified more quickly (that is, at smaller depths of continuation).

A7.3 Polynomial expression

It is evident from the form of (A 7.1) and (A 7.2) that for numerical evaluation we can replace a Fourier transform integral by a polynomial in the variable $z = \exp(-i2\pi v \Delta t)$, and the inverse Fourier transform by one in $1/z = \exp(i2\pi v \Delta t)$. Thus, for example,

$$F(v) = \sum_n A_n z^n$$

where the polynomial coefficients A_n are given by the integral of $f(t)$ over the nth Δt-interval.

References

[1] STACEY, F. D. and BANERJEE, S. K. (1974). *The Physical Principles of Rock Magnetism*. (Elsevier, Amsterdam).
[2] GREEN, R. (1960). *Geophys. Prosp.,* **8**, 98–110.
[3] GEYGER, W. A. (1962). *AIME Trans.*, **81**, 65–73.
[4] GORDON, D. I. *et al*. (1968). *IEEE Trans. Mag.*, **Mag.-4**, 397–401.
[5] PRIMDAHL, F. (1970). *IEEE Trans. Mag.*, **Mag.-6**, 376–83.
[6] GORDON, D. I. *et al*. (1972). *IEEE Trans. Mag.*, **Mag.-8**, 76–82.
[7] WATERS, G. S. *et al*. (1965). *Geophys. Prosp.,* **4**, 1–9.
[8] GUPTA SARMA, D. *et al*. (1966). *Geophys. Prosp.,* **14**, 292–300.
[9] GIRET, R. *et al*. (1965). *Geophys. Prosp.,* **13**, 225–39.
[10] ROY, A. (1970). *Geoexploration,* **8**, 37–40.
[11] LEHMANN, H. (1971). *Geophys. Prosp.,* **19**, 133–55.
[12] HAALCK, H. (editor) (1952). *Lehrbuch d. angew. Geoph.*, Teil 1, (Borntraeger, Berlin).
[13] CARLHEIM-GYLLENSKÖLD, V. (1910). *A Brief Account of a Magnetic Survey of the Iron Ore Field at Kirunavaara*. (Stockholm).
[14] ÅM, K. (1972). *Geoexploration,* **10**, 63–90.
[15] HOOD, P. (1964). *Geophys. Prosp.,* **12**, 440–456.
[16] GAY, S. P. (1963). *Geophysics,* **28**, 161–200.
[17] HUTCHISON, R. (1958). *Geophysics,* **23**, 749–69.
[18] HJELT, S. E. (1972). *Geoexploration,* **10**, 239–54.
[19] JOSEPH, R. I. *et al*. (1965). *J. Appl. Phys.*, **36**, 1579–93.
[20] JOSEPH, R. I. (1976). *Geophysics*, **41**, 1052–4.
[21] SHUEY, R. T. *et al*. (1972). *Geoexploration*, **10**, 221–8.
[22] BOTT, M. (1963). *Geophys. Prosp.,* **9**, 292–9.
[23] SHARMA, P. V. (1967). *Geophys. Prosp.,* **15**, 167–73.
[24] SMITH, R. A. (1959). *Geophys. Prosp.,* **7**, 55–63.
[25] SMITH, R. A. (1961). *Geophys. Prosp.,* **3**, 399–410.
[26] YÜNGUL, S. (1956). *Geophysics,* **21**, 433–54.
[27] ZACHOS, K. (1944). *Beitr. angew. Geophys.,* **11**, 1.
[28] WERNER, S. (1945). *Sveriges Geologiska Undersökning (Stockholm)*, **39**(5), 1–79.

262

[29] MOONEY, H. (1952). *Geophysics*, **17**, 531–43.
[30] MALMQVIST, D. (1957). *Freiberger Forschungshefte*, **C 32**, 20–39.
[31] COLLINSON, D. W., CREER, K. M. and RUNCORN, S. K. (1967). *Methods in Palaeomagnetism.* (Elsevier, Amsterdam).
[32] Geodetic reference system 1967, *Publ. Spec. du Bulletin Géodésique.* (Bureau Central de L'Association Internationale de Géodesie, Paris).
[33] COOK, A. H. (1973). *Physics of the Earth and Planets* (Macmillan, London).
[34] HEILAND, C. A. (1940). *Geophysical Exploration.* (Prentice Hall, New York).
[35] LACOSTE, L. J. B. (1934). *Physics*, **5**, 178–80.
[36] MELTON, B. S. (1971). *Geophys. J.*, **22**, 521.
[37] SIIKARLA, T. (1966). *Geoexploration*, **4**, 139–49.
[38] KETELAAR, A. C. R. (1976). *Geoexploration*, **14**, 57–65.
[39] *Tidal Gravity Corrections for 1978.* (European Association of Exploration Geophysicists, The Hague).
[40] DEHLINGER, P. *et al.* (1972). *Marine Geology*, **12**, 1–41.
[41] NETTLETON, L. L. (1939). *Geophysics*, **4**, 176–83.
[42] JUNG, K. (1953). *Zeit. f. Geoph.*, **19**, *Sonderband*, 54–8.
[43] PARASNIS, D. S. (1952). *Month. Not. Roy. Astr. Soc., Geophys. Suppl.*, **6**, 252.
[44] SIEGERT, A. J. F. (1942). *Geophysics*, **7**, 29–34.
[45] LEGGE, J. A. (1944). *Geophysics*, **9**, 175–9.
[46] JUNG, K. (1959). *Gerland Beitr. Geoph.*, **68**, 268–79.
[47] SYBERG, F. J. R. (1972). *Geophys. Prosp.*, **20**, 47–75.
[48] SINGH, S. K. (1977). *Geophys. J. Roy. Astr. Soc.*, **50**, 243–6.
[49] LINDBLAD, A. *et al.* (1938). *Ingenjörs Vet. Akad. Handl. No. 146*, 52.
[50] LEVINE, S. (1941). *Geophysics*, **6**, 180–96.
[51] MORGAN, N. A. *et al.* (1972). *Geophys. Prosp.*, **20**, 363–74.
[52] TALWANI, M. *et al.* (1960). *Geophysics*, **25**, 203–25.
[53] SKEELS, D. C. (1947). *Geophysics*, **12**, 43–56.
[54] AL-CHALABI, M. (1972). *Geophys. Prosp.*, **20**, 1–16.
[55] ROY, A. (1966). *Geoexploration*, **4**, 65–83.
[56] JUNG, K. (1937). *Zeit. f. Geoph.*, **13**, 45.
[57] JUNG, K. (1953). *Geophys. Prosp.*, **1**, 29–35.
[58] BOTT, M. H. P. *et al.* (1958). *Geophys. Prosp.*, **6**, 1–10.
[59] SMITH, R. A. (1959). *Geophys. Prosp.*, **7**, 55–63.
[60] SMITH, R. A. (1960). *Geophys. Prosp.*, **8**, 607–13.
[61] GRANT, F. and WEST, G. (1965). *Interpretation Theory in Applied Geophysics* (McGraw-Hill, New York).
[62] COOK, A. H. *et al.* (1951). *Quart. J. Geol. Soc. Lond.*, **107**, 287–306.
[63] DAVIS, W. E. *et al.* (1957). *Geophysics*, **22**, 848–69.
[64] EVE, A. S. and KEYS, D. A. (1956). *Applied Geophysics*, 4th ed. (Cambridge University Press, Cambridge).

[65] KOLBENHEYER, T. *et al.* (1969). *Geoexploration*, 7, 153–62.
[66] PARASNIS, D. S. (1975), *Mining Geophysics*, 2nd ed. (Elsevier, Amsterdam).
[67] GAY, S. P. (1967). *Geophys. Prosp.*, **15**, 236–45.
[68] OGILVY, A. A. *et al.* (1969). *Geophys. Prosp.*, **17**, 36–62.
[69] SATO, M. and MOONEY, H. M. (1960). *Geophysics*, **25**, 226–49.
[70] MUSKAT, M. and EVINGER, H. H. (1941). *Geophysics*, **6**, 397–427.
[71] CARPENTER, E. W. *et al.* (1956). *Geophysics*, **21**, 455–69.
[72] PATELLA, D. (1974). *Geophys. Prosp.*, **22**, 315–29.
[73] SLICHTER, L. B. (1933). *Physics*, **4**, 307–22.
[74] LANGER, R. E. (1933). *Amer. Math. Soc. Bull.*, **29**, 814–20.
[75] STEVENSON, A. F. (1934). *Physics*, **5**, 114–24.
[76] HUMMEL, J. N. (1929). *Zeit. f. Geoph.*, **5**, 89.
[77] HUMMEL, J. N. (1929). *Zeit. f. Geoph.*, **5**, 228.
[78] *Standard graphs for resistivity prospecting* (1969). (European Association of Exploration Geophysicists, The Hague).
[79] MOONEY, H. M. and WETZEL, W. W. (1956). *The Potentials about a Point Electrode.* (University of Minnesota Press, Minneapolis).
[80] BHATTACHARYA, P. K. and PATRA, H. P. (1968). *Direct current electrical sounding.* (Elsevier, Amsterdam).
[81] ZOHDY, A. A. R. (1965). *Geophysics*, **30**, 644–60.
[82] PEKERIS, C. L. (1940). *Geophysics*, **5**, 31–42.
[83] KOFOED, O. (1968). *The Application of the Kernel Function in Interpreting Geoelectrical Resistivity Measurements.* (Gebrüder Borntraeger, Berlin).
[84] PATELLA, D. (1975). *Geophys. Prosp.*, **23**, 335–62.
[85] GHOSH, D. P. (1971). *Geophys. Prosp.*, **19**, 769–75.
[86] JOHANSEN, H. K. (1975). *Geophys. Prosp.*, **23**, 449–58.
[87] INMAN, J. R. *et al.* (1973). *Geophysics*, **38**, 1088–108.
[88] JOHANSEN, H. K. (1977). *Geophys. Prosp.* (in press).
[89] FLATHE, H. (1955). *Geophys. Prosp.*, **3**, 95–110.
[90] ORELLANA, E. (1963). *Geophysics*, **28**, 199–210.
[91] MAILLET, R. (1947). *Geophysics*, **12**, 529–56.
[92] LOGN, O. (1954). *Geophysics*, **19**, 739–60.
[93] MAEDA, K. (1955). *Geophysics*, **20**, 123–47.
[94] DE GERY, J. C. *et al.* (1956). *Geophysics*, **21**, 780–93.
[95] LEE, T. (1972). *Geophys. Prosp.*, **20**, 847–59.
[96] COOK, K. L. *et al.* (1954). *Geophysics*, **19**, 761–90.
[97] ALFANO, L. (1959). *Geophys. Prosp.*, **7**, 311–66.
[98] VOZOFF, K. (1958). *Geophysics*, **23**, 536–56.
[99] COLOMBO, U. *et al.* (1959). *Geophysics*, **7**, 91–118.
[100] LYNCH, E. J. (1962). *Formation Evaluation.* (Harper and Row, New York).
[101] SHUEY, R. T. (1975). *Semi-conducting ore minerals.* (Elsevier, Amsterdam).

[102] PARKHOMENKO, E. (1967). *Electrical Properties of Rocks*. (Plenum Press, New York).
[103] HARTSHORN, L. (1925). *J. Inst. elect. Engrs.*, **64**, 1152.
[104] BERTIN, J. and LOEB, J. (1976). *Experimental and Theoretical Aspects of Induced Polarization*, Vols 1 and 2. (Gebrüder Borntraeger, Berlin).
[105] SUMNER, J. S. (1976). *Principle of induced polarization for geophysical exploration*. (Elsevier, Amsterdam).
[106] WAIT, J. (editor) (1959). *Overvoltage Research and Geophysical Applications*. (Pergamon, New York) p. 158.
[107] NILSSON, B. (1971). *Geoexploration*, **9**, 35–54.
[108] HOHMAN, G. W. (1973). *Geophysics*, **38**, 854–63.
[109] RATHOR, B. S. (1977). *Geoexploration*, **15**, 137–49.
[110] WYNN, J. C. *et al.* (1977). *Geophys. Prosp.*, **25**, 29–51.
[111] PATELLA, D. (1973). *Geophys. Prosp.*, **21**, 315–29.
[112] HALLOF, P. (1967). *IP newsletter cases* xxvii *and* xxviii. (McPhar Geophysics Ltd, Canada).
[113] PARASNIS, D. S. (1966). *Geoexploration*, **4**, 177–208.
[114] SUNDBERG, K. (1931). *Gerlands Beitr. Geoph., Ergänzungs*-Hefte, **1**, 298–361.
[115] LEVI-CIVITA (1902). *Atti della Reale Accademia dei Lincei*.
[116] SUNDBERG, K. *et al.* (1934). *Structural investigations by electromagnetic methods, World Petrol. Congr., B(1)*, 107–18.
[117] HEDSTRÖM, H. (1930). *The Oil Weekly*, July 25 and August 8 issues.
[118] HEDSTRÖM, H. (1937). *Amer. Inst. Min. Met. Eng.*, Tech. Publ. 827.
[119] BEZVODA, V. *et al.* (1970). *Geophys. Prosp.*, **3**, 343–51.
[120] DIZIOGLU, M. Y. (1967). *Geoexploration*, **5**, 157–64.
[121] *Geophysical Surveys in Mining, Hydrological and Engineering Projects* (1958). (European Association of Exploration Geophysicists, The Hague).
[122] PARASNIS, D. S. (1971). *Geophys. Prosp.*, **19**, 163–79.
[123] NAIR, M. R. *et al.* (1968). *Geoexploration*, **6**, 207–44.
[124] PAÁL, G. (1965). *Geoexploration*, **3**, 139–47.
[125] PATERSON, N. *et al.* (1971). *Geoexploration*, **9**, 7–26.
[126] PHILIPS, W. J. *et al.* (1975). *Geoexploration*, **13**, 215–26.
[127] NORTON, K. A. (1937). *Proc. I.R.E.*, **25**, 1203–36.
[128] DEBYE, P. (1909). *Ann. d. Phys.*, **30**, 59.
[129] MARCH, H. W. (1953). *Geophysics*, **16**, 671–84.
[130] WAIT, J. R. (1960). *Geophysics*, **25**, 649–58.
[131] MALMQVIST, D. (1965). *Geoexploration*, **3**, 175–227.
[132] WAIT, J. R. (1952). *Geophysics*, **17**, 378–86.
[133] WESLEY, J. P. (1958). *Geophysics*, **23**, 134–43.
[134] GRAF, A. (1934). *Gerlands Beitr. Geoph.*, **4**, 1.
[135] SLICHTER, L. B. (1951). *Geophysics*, **16**, 431–49.
[136] SLICHTER, L. B. *et al.* (1959). *Geophysics*, **24**, 77–88.

[137] NEGI, J. G. (1967). *Geophysics*, **32**, 69–87.
[138] LAJOIE, J. J. *et al.* (1976). *Geophysics*, **41**, 1133–56.
[139] OLSSON, O. (1978). *Radio Science*, **13**, No. 2.
[140] WARD, S. H. (1971). *Geophysics*, **36**, 1–183.
[141] WARD, S. H. (1976). *Geophysics*, **41**, 1103–258.
[142] KOFOED, O. *et al.* (1972). *Geophys. Prosp.*, **20**, 406–20.
[143] JOHANSEN, H. K. (1976). *Geophys. Prosp.*, **24**, 605–06.
[144] HARRISON, C. H. (1970). *Geophysics*, **35**, 1099–115.
[145] YOST, W. J. (1952). *Geophysics*, **17**, 89–106.
[146] MORRISON, H. F. *et al.* (1969). *Geophys. Prosp.*, **17**, 82–101.
[147] MORLEY, L. W. (editor) (1970). *Mining and Groundwater Geophysics* (1967). (Department of Energy, Mines and Resources, Ottawa).
[148] PARASNIS, D. S. (1974). *Geoexploration*, **12**, 97–120.
[149] VERMA, S. K. (1975). *Geophys. Prosp.*, **23**, 292–9.
[150] NABIGHIAN, M. (1971). *Geophysics*, **36**, 25–37.
[151] HJELT, S. E. (1971). *Geoexploration*, **9**, 213–30.
[152] HURLEY, D. G. (1977). *Geoexploration*, **15**, 77–85.
[153] LEE, T. *et al.* (1974). *Geophys. Prosp.*, **22**, 430–44.
[154] SINGH, R. N. (1972). *Geophysics*, **38**, 864–93.
[155] LEE, T. (1975). *Geophys. Prosp.*, **23**, 492–512.
[156] TUMAN, V. S. (1951). *Geophysics*, **16**, 102–14.
[157] NIBLETT, E. R. *et al.* (1960). *Geophysics*, **25**, 998–1008.
[158] PORSTENDORFER, G. (1961). *Geophys. Prosp.*, **9**, 128–43.
[159] YÜNGUL, S. (1977). *Geoexploration*, **15**, 207–38.
[160] CAGNIARD, L. (1953). *Geophysics*, **18**, 605–35.
[161] VOZOFF, K. (1970). *Mining and Groundwater Geophysics* (1967). Edited by L. W. Morley (Department of Energy, Mines, and Resources, Ottawa).
[162] KELLER, G. (1971). *Geoexploration*, **9**, 99–148.
[163] PORSTENDORFER, G. (1974). *Principles of Magnetotelluric Prospecting.* (Gebrüder Borntraeger, Berlin).
[164] WARD, S. H. (1961). *Geophys. Prosp.*, **9**, 191–202.
[165] HEDSTRÖM, H. *et al.* (1959). *Geophys. Prosp.*, **7**, 448–70.
[166] PATERSON, N. R. (1961). *Geophysics*, **26**, 601–47.
[167] TÖRNQUIST, G. (1958). *Geophys. Prosp.*, **6**, 112–26.
[168] HEDSTRÖM, H. *et al.* (1958). *Geophys. Prosp.*, **6**, 448–70.
[169] MAKOWIECKI, L. Z. *et al.* (1970). *Inst. Geol. Sci., Geophys. Paper No. 3* (London).
[170] WARD, S. H. (1959). *Geophysics*, **24**, 671–89.
[171] GUPTA SARMA, D. *et al.* (1976). *Geophysics*, **41**, 287–99.
[172] VERMA, S. K. (1972). *Geophys. Prosp.*, **20**, 752–70.
[173] BECKER, A. (1967). *Geoexploration*, **5**, 81–8.
[174] RICKER, N. (1953). *Geophysics*, **18**, 10–40.

[175] RICKER, N. H. (1977). *Transient waves in visco-elastic media*. (Elsevier, Amsterdam).

[176] FAUST, L. Y. (1951). *Geophysics*, 16, 192–206.

[177] HUGHES, D. S. *et al.* (1951). *Geophysics*, 16, 577–93.

[178] BAULE, H. (1953). *Geophys. Prosp.*, 1, 111–24.

[179] SHUMWAY, G. (1956). *Geophysics*, 21, 305–19.

[180] DOBRIN, M. (1960). *Introduction to Geophysical Prospecting*. (McGraw-Hill, New York).

[181] EVENDEN, B. S. and STONE, D. R. (1971). *Seismic Prospecting Instruments*, Vol. 2. (Gebrüder Borntraeger, Berlin).

[182] MOTA, L. (1954). *Geophysics*, 19, 242–54.

[183] JOHNSON, S. H. (1976). *Geophysics*, 41, 418–24.

[184] MEIDAV, T. (1960). *Geophysics*, 25, 1035–53.

[185] HABBERJAM, G. (1966). *Geoexploration*, 4, 219–25.

[186] *Geophysical surveys in Mining, Hydrological and Engineering Projects* (1958). (European Association of Exploration Geophysicists, The Hague).

[187] GREEN, R. (1974). *Geoexploration*, 12, 259–84.

[188] KREY, T. (1951). *Geophysics*, 16, 468–85.

[189] KREY, T. (1954). *Geophysics*, 2, 61–72.

[190] LORENZ, G. (1954). *Gerlands Beitr. Geophys.*, 63, 99.

[191] BAUMGARTE, J. (1955). *Geophys. Prosp.*, 3, 126–62.

[192] TELFORD, W. M., GELDART, T. M., SHERIFF, R. E. and KEY, D. A. (1976). *Applied Geophysics*. (Cambridge University Press, Cambridge).

[193] MENZEL, H. (1957). *Geophys. Prosp.*, 5, 328–48.

[194] Symposium (1958). *Geophys. Prosp.*, 6, 394–455.

[195] ANSTEY, N. A. (1957). *Geophys. Prosp.*, 5, 44–68.

[196] MUSKAT, M. *et al.* (1940). *Geophysics*, 5, 115–48.

[197] BERRYMAN, L. H. *et al.* (1958). *Geophysics*, 23, 223–52.

[198] BORN, W. T. (1941). *Geophysics*, 6, 132–48.

[199] DATTA, S. (1968). *Geoexploration*, 6, 127–39.

[200] SMITH, M. K. (1958). *Geophysics*, 23, 44–57.

[201] MAYNE, H. (1962). *Geophysics*, 27, 927–38.

[202] PETERSON, R. A. *et al.* (1955). *Geophysics*, 20, 516–38.

[203] Symposium on synthetic seismograms (1960). *Geophys. Prosp.*, 8, 231–346.

[204] ROBINSON, E. A. (1966). *Geophys. Prosp.*, 14, Suppl 1.

[205] D'HOERAENE, J. (1960). *Geophys. Prosp.*, 8, 389–400.

[206] KUNETZ, G. (1961). *Geophys. Prosp.*, 9, 317–41.

[207] RICE, R. B. (1962). *Geophysics*, 27, 4–18.

[208] AGUIELERA, R. *et al.* (1970). *Geophysics*, 35, 247–53.

[209] VOGEL, C. B. (1952). *Geophysics*, 17, 586–97.

[210] SUMMERS, G. C. *et al.* (1952). *Geophysics*, 17, 598–614.

[211] KREY, T. (1969). *Geophys. Prosp.*, **17**, 206–18.
[212] EDELMAN, H. (1966). *Geophys. Prosp.*, **14**, 455–69.
[213] BARBIER, M. *et al.* (1970). *Geophys. Prosp.*, **18**, 571–80.
[214] MOXHAM, R. M. (1960). *Geophysics*, **25**, 408–32.
[215] ZESCHKE, G. (1963). *Econ. Geol.*, **58**, 995–6.
[216] BUDDE, E. (1958). *Geophys. Prosp.*, **6**, 25–34.
[217] HOMILIUS, J. *et al.* (1957). *Geophys. Prosp.*, **5**, 449–68.
[218] HOMILIUS, J. *et al.* (1958). *Geophys. Prosp.*, **6**, 342–64.
[219] SCHWARZER, T. F. *et al.* (1973). *Econ. Geol.*, **68**, 1297–312.
[220] BRODING, R. A. *et al.* (1955). *Geophysics*, **20**, 841–59.
[221] HÄHNEL, R. and KAPPELMEYER, O. (1974). *Geothermics with special reference to application.* (Gebrüder Borntraeger, Berlin).
[222] POOLEY, J. Ph. *et al.* (1970). *Geophys. Prosp.*, **18**, 666–700.
[223] GINZBURG, I. I. (1960)., *Principles of geochemical prospecting.* (Pergamon Press, New York).
[224] HAWKES, H. E. and WEBB, J. S. (1962). *Geochemistry in mineral exploration.* (Harper, New York).
[225] KENDALL, M. G. and MORAN, P. A. P. (1963). *Geometrical Probability.* (Charles Griffin and Co., London).
[226] AGOCS, W. B. (1955). *Geophysics*, **20**, 871–85.

Index